Aphid Biodiversity under Environmental Change

Pavel Kindlmann · A.F.G. Dixon ·
J.P. Michaud
Editors

Aphid Biodiversity under Environmental Change

Patterns and Processes

Springer

Editors
Prof. Pavel Kindlmann
Institute of Systems Biology
and Ecology AS CR and
Institute for Environmental Studies
Charles University
CZ-128 01 Prague 2
Czech Republic
pavel.kindlmann@centrum.cz

Prof. A.F.G. Dixon
Institute of Systems Biology
and Ecology AS CR and
School of Biological Sciences
University of East Anglia
NR4 7TJ Norwich
United Kingdom
a.f.dixon@uea.ac.uk

Prof. J.P. Michaud
Department of Entomology
Kansas State University
240th Ave. 1232
67601 Hays Kansas
USA
jpmi@ksu.edu

ISBN 978-90-481-8600-6 e-ISBN 978-90-481-8601-3
DOI 10.1007/978-90-481-8601-3
Springer Dordrecht Heidelberg London New York

Library of Congress Control Number: 2010929761

Publication of this book was supported by the Biodiversity Research Centre of the CR and by Center of Excellence CZECHGLOBE.

Printed on acid-free paper

Springer is part of Springer Science+Business Media (www.springer.com)

Preface

This book presents the results of recent research on aphid population dynamics and ecology relevant to current environmental changes resulting from global warming. It incorporates a selection of the contributions presented at the International Symposium on Aphids in Fremantle, Australia, in October 2005, plus some additional invited chapters. The objective was to incorporate the major issues in the field and simultaneously create a closely interrelated and integrated volume.

The first chapter sets the scene. Kindlmann and Dixon present a critical review of existing models of aphid population dynamics, examine the biological assumptions that are incorporated in the models and present one of the latest models of aphid metapopulation dynamics. They conclude that natural enemies are unlikely to affect aphid population dynamics late in a season, but in some years may have an effect very early in the season, when aphid colonies are still small and predators might be able to reduce the numbers of colonies.

The question, whether aphids will move to different locations, adapt to the change in conditions in their current habitat or go extinct is discussed by Ameixa. She concludes that the distributions of aphids are most likely to change, with the distribution of each species moving globally as their preferred habitat moves in response to changes in the climate, which may be more difficult than in the past because of habitat fragmentation and habitat loss.

The chapter by Harrington and Clark makes use of suction trap catches, the best available long-term data in the world on aphid dynamics. Acknowledging the difficulties of interpretation and prediction, they investigate whether there is any evidence that the start of the spring flight of aphids at particular locations is occurring earlier as winters get warmer, and whether this is more so for aphids that tend to be continuously parthenogenetic at that locality, and that the trend in the migration beginning earlier and ending later in more recent years is a consequence of an increase in average temperatures.

The paper by van Baaren et al. addresses the effects of climate on insect communities, focusing on aphids, aphid parasitoids, predators and hyperparasitoids. For each trophic level, the general effect of temperature change on insects is discussed, with emphasis on species belonging to aphid-based communities.

Michaud in his chapter claims that an increase in annual mean temperatures would have certain predictable consequences for cereal aphid populations via direct effects on aphid biology, and indirectly through effects on plants and natural enemies. However, any sustained shift in prevailing wind patterns associated with atmospheric warming could generate an unpredictable cascade of ecological consequences for both agriculture and cereal aphids, mediated largely by changes in rainfall patterns and migration pathways, respectively.

Qureshi considers the impact of increasing temperature on the aphid species *Toxoptera citricida*, a cosmopolitan pest of citrus and a highly efficient vector of citrus tristeza virus. He concludes that the negative impact of rising temperatures on *T. citricida* populations may be more pronounced in the south than elsewhere due to the relatively higher temperatures there, with more beneficial effects evident in the north.

Clement et al. investigate the history of pea aphid outbreaks in the U.S. Pacific Northwest. Various abiotic and biotic factors and their possible controlling influence on changes in pea aphid densities, with emphasis on winter temperatures within the context of climate change, are presented and discussed.

Dixon and Hopkins study the mechanisms of coexistence of several aphid species on the same host plant. Using data on five species of aphids coexisting on the leaves of birch and indicate that the temporal patterns in their reproductive activity are associated with differences in their thermal tolerances.

Roy and Majerus deal with the role of ladybirds in the changing world. They conclude that it is difficult to assess the impact that anthropogenic factors will have on most species of coccinellid but logic suggests that the direct anthropogenic drivers of environmental change, both individually and in concert, will be highly deleterious to all but the most adaptable and eurytopic coccinellids.

Aphids have evolved a particular form of inducible anti-predator behaviour that involves the emission of alarm pheromone. Outreman et al. show that alarm signaling in aphids is associated with the ecological cost of attracting additional natural enemies and demonstrate that a full understanding of the evolution of inducible defenses has to consider a species' complete network of ecological interactions.

This book fills a significant gap in the recent literature: while there are several books on aphid biology and ecology and their importance as crop pests (Dixon 1998, 2005; Minks and Harrewijn 1987–1989; van Emden and Harrington 2007) there are none on the possible effects of environmental changes on aphid population dynamics and their biodiversity. This book will be a useful introduction to the subject for graduate students, researchers in crop science, crop protection, agricultural advisors and managers etc., but can also be used as a complementary text in any course on population dynamics and ecology of crop pests for undergraduates or graduates.

České Budějovice, Czech Republic — Pavel Kindlmann
Norwich, UK — A.F.G. Dixon
Manhattan, Kansas — J.P. Michaud
November 2009

Contents

Contributors

Jacques JM van Alphen UMR 6553 ECOBIO, Université de Rennes I, 35 042 Rennes Cedex, France; Van der Klaauw Laboratory, Institute of Biology Leiden, 2300 RA Leiden, The Netherlands

Olga M.C.C. Ameixa Department of Theoretical Ecology, Institute of Systems Biology and Ecology AS CR, CZ-370 05 České Budějovice, Czech Republic, olgameixa@portugalmail.pt

Joan van Baaren UMR – CNRS 6553 ECOBIO, Université de Rennes I, 35 042 Rennes Cedex, France, joan.van-baaren@univ-rennes1.fr

Suzanne Clark Rothamsted Research, Hertfordshire AL5 2JQ, UK

Stephen L. Clement USDA-ARS, Plant Germplasm Introduction and Testing Research Unit, Washington State University, Pullman, WA 99164-6402, USA, slclement@wsu.edu

Anthony F.G. Dixon Department of Theoretical Ecology, Institute of Systems Biology and Ecology AS CR, CZ-370 05 České Budějovice, Czech Republic; School of Biological Sciences, University of East Anglia, Norwich NR4 7TJ, UK, a.f.dixon@uea.ac.uk

Sanford D. Eigenbrode Department of Plant, Soil and Entomological Sciences, University of Idaho, Moscow, ID 83844-2339, USA

Richard Harrington Rothamsted Research, Hertfordshire AL5 2JQ, UK, richard.harrington@bbsrc.ac.uk

Graham W. Hopkins Invertebrate Surveys Ltd, Norwich NR1 1ES, UK

Damon S. Husebye Department of Plant, Soil and Entomological Sciences, University of Idaho, Moscow, ID 83844-2339, USA

Pavel Kindlmann Department of Theoretical Ecology, Institute of Systems Biology and Ecology AS CR, CZ-370 05 České Budějovice, Czech Republic; Faculty of Science, Institute for Environmental Studies, Charles University, CZ-128 01 Prague 2, Czech Republic, pavel.kindlmann@centrum.cz

Grit Kunert UMR INRA-Agrocampus Rennes "Biologie des Organismes et des Populations appliquée à la Protection des Plantes" [BIO3P], Le Rheu Cedex, France; Department of Biochemistry, Max-Planck-Institute for Chemical Ecology, 07745 Jena, Germany

Cécile Le Lann UMR 6553 ECOBIO, Université de Rennes I, 35 042 Rennes Cedex, France

Michael E.N. Majerus Department of Genetics, Cambridge University, Cambridge, CB2 3EH, UK

J.P. Michaud Department of Entomology, Kansas State University, Manhattan, KS, USA, jpmi@ksu.edu

Yannick Outreman UMR 1099 INRA-Agrocampus Ouest-Université Rennes 1 "Biologie des Organismes et des Populations appliquée à la Protection des Plantes" [BIO3P], Le Rheu Cedex, France; Institute of Ecology, Friedrich-Schiller-University, 07743 Jena, Germany, yannick.outreman@rennes.inra.fr

Jawwad A. Qureshi Department of Entomology and Nematology, Institute of Food and Agricultural Sciences, University of Florida, Immokalee, FL 34142, USA, jawwadq@ufl.edu

Helen E. Roy NERC Centre for Ecology and Hydrology, Wallingford, Oxfordshire OX10 8BB, UK, hele@ceh.ac.uk

Jean-Christophe Simon UMR 1099 INRA-Agrocampus Ouest-Université Rennes 1 "Biologie des Organismes et des Populations appliquée à la Protection des Plantes" [BIO3P], Le Rheu Cedex, France

Wolfgang W. Weisser Institute of Ecology, Friedrich-Schiller-University, 07743 Jena, Germany

Chapter 1
Modelling Population Dynamics of Aphids and Their Natural Enemies

Pavel Kindlmann and Anthony F.G. Dixon

Abstract Aphids are serious pests of many agricultural crops. Therefore, a good understanding of their population dynamics is vitally important for crop protection. There have been several attempts made to forecast the abundance of aphids and develop expert systems to help farmers optimize prophylactic measures and minimize their costs. The advisory systems, however, did not receive general acceptance and disappointingly few forecasting systems are in use. The failure of models to predict aphid population dynamics for practical purposes is due to the extremely wild oscillations in aphid numbers caused by intrinsic (size, fecundity, mortality, migration rate) and external factors (weather, especially temperature). As a consequence, the predictions are unlikely to be robust enough for reliable forecasting, mainly because they depend on the course of weather during the season, which cannot be predicted. Here we present a critical review of existing models of aphid population dynamics, examine biological assumptions that are incorporated in the models and present one of the latest models of aphid metapopulation dynamics. We conclude that natural enemies are unlikely to affect aphid population dynamics late in the season, but may have an effect very early in the season, when aphid colonies are still small and predators might be able to reduce the numbers of these colonies. Empirical verification of this is still very weak, however, and further experiments on this aspect of predator prey dynamics should be undertaken.

Keywords Aphids · Aphidophagous insects · Population dynamics · Predictive models · Trophic interactions

---

P. Kindlmann (✉)
Department of Theoretical Ecology, Institute of Systems Biology and Ecology AS CR, CZ-370 05 České Budějovice, Czech Republic; Faculty of Science, Institute for Environmental Studies, Charles University, CZ-128 01 Prague 2, Czech Republic
e-mail: pavel.kindlmann@centrum.cz

P. Kindlmann et al. (eds.), *Aphid Biodiversity under Environmental Change*,
DOI 10.1007/978-90-481-8601-3_1,

1.1 Introduction

Aphids are serious pests of many agricultural crops. Therefore, a good understanding of their population dynamics is vitally important for crop protection. Not surprisingly, the pest status of aphids and political concern over the prophylactic application of pesticides attracted the attention of modellers starting in the 1960s (Hughes 1963; Hughes and Gilbert 1968; Gilbert and Hughes 1971; Gosselke et al. 2001). Attempts were made to forecast the abundance of aphids and propose expert systems to help farmers optimize prophylactic measures and minimize their costs (Mann et al. 1986; Gonzalez-Andujara et al. 1993; Ro and Long 1999). These studies usually concluded that forecasting is a better strategy than either no control or prophylaxis, where yields are average and above (Watt 1983; Watt et al. 1984). The advisory systems, however, did not receive general acceptance and disappointingly few forecasting systems are in use. Analysis of some of the existing models of aphid population dynamics reveals the reasons. For example, a model that describes the summer population dynamics of the grain aphid, *Sitobion avenae* (Carter et al. 1982; Carter 1985) was modified and extended to include the population dynamics of the aphidophagous predator *Coccinella septempunctata* (Skirvin et al. 1997a,1997b). It is claimed to give better predictions than the Carter et al. (1982) model, but there are few data against which it can be validated. The main weakness of the Skirvin et al. (1997a) model is that it gives the same prediction for identical initial conditions, which is contrary to what is observed in the field.

Early models of the population dynamics of the peach–potato aphid, *Myzus persicae* (Scopes 1969; Tamaki and Weeks 1972, 1973; Tamaki 1973, 1984; DeLoach 1974; Taylor 1977; Whalon and Smilowitz 1979; Tamaki et al. 1980, 1982; Mack and Smilowitz 1981, 1982; Smilowitz 1984; Ro and Long 1998) were recently improved by Ro and Long (1999). However, even this model is not validated against data that were not used to derive the parameters, which devalues the claim that it gives a good prediction. In addition, it also makes the unwarranted assumption that the decline in aphid abundance is caused by predators (see later in this chapter).

A simulation model developed to investigate the interrelationship of factors influencing the population dynamics of the bird cherry-oat aphid, *Rhopalosiphum padi*, in barley crops during autumn and winter (Morgan 2000) accurately predicts outbreaks and the peak aphid populations within 20% of that observed in all but one case. However, this model is not suitable for long-term predictions, as it requires the daily input of maximum and minimum temperatures, which invalidates its predictive value, as these temperatures cannot be predicted with sufficient accuracy. Another model for this species was developed by Wiktelius and Pettersson (1985), but was not used for forecasting and therefore there is a need for further research on this topic.

A whole family of models of the cowpea aphid, *Aphis craccivora* (Gutierrez et al. 1974), and the pea aphid, *Acyrthosiphum pisum*, population dynamics (Gutierrez and Baumgärtner 1984a, 1984b; Gutierrez et al. 1984), and that of their natural enemies (Gutierrez et al. 1980, 1981) were developed by Gutierrez and his group, but even these were not used for long-term predictions. Similarly, a computer simulation model developed to investigate spatial and population dynamics of apterae of the Russian wheat aphid, *Diuraphis noxia* on preferred (wheat) and non-preferred (oat) hosts by Knudsen and Schotzko (1991) is only suitable for short-term (14 and 21 days) predictions. A transition matrix model developed to simulate the population dynamics of the green apple aphid, *Aphis pomi* (Woolhouse and Harmsen 1991) has also not been validated against an independent data set.

Recently, spatio-temporal, or metapopulation models have been published (Weisser 2000; Winder et al. 2001). These are a promising development, but modellers employing this approach need to consider whether aphid migration, rather than predator-inflicted mortality, is the regulating factor. The question remains, whether predators drive aphid metapopulation dynamics, or as predicted by theory (Kindlmann and Dixon 1996, 1999) are responding to aphid abundance, which is self-regulated by migration.

In general, the failure of models to predict aphid population dynamics for practical purposes is due to the extremely wild oscillations in aphid numbers caused by intrinsic factors (size, fecundity, mortality, migration rate) and external factors (weather, especially temperature). As a consequence, the predictions are unlikely to be robust enough for reliable forecasting, mainly because they depend on the course of weather during the season, which cannot be predicted. In addition, most of the models tend to be very complex, which stems from the belief of their authors that complexity means better accuracy, which is not always the case (Stewart and Dixon 1988). This is because the measuring errors, associated with each of the large number of parameters, yield highly variable predictions. Thus, there is a serious gap in our knowledge, which needs to be filled in order to confirm or refute the understanding arrived at mainly by studying aphids living on woody plants. For a further discussion of forecasting, see Harrington et al. (2007).

1.2 Biological Background

1.2.1 Aphid Biology Relevant to Population Dynamics

Most aphid species can reproduce both asexually and sexually, with several parthenogenetic generations between each period of sexual reproduction. This is known as cyclical parthenogenesis and, in temperate regions, sexual reproduction occurs in autumn and results in the production of overwintering eggs, which hatch the following spring and initiate another cycle. Many pest aphids, however, do not overwinter as an egg but as nymphs or adults and others as both eggs and active stages (see Williams and Dixon 2007). For their size, the parthenogenetic individuals have very short developmental times and potentially prodigious rates of increase (de Réaumur 1737; Huxley 1858; Kindlmann and Dixon 1989; Dixon 1992). Thus, aphids show very complex and rapidly changing within-year dynamics, with each clone going through several generations during the vegetative season and being made up of many individuals, which can be widely scattered in space. The survival of the eggs and/or overwintering aphids determines the numbers of aphids present the following spring.

The study of the population dynamics of aphids living on herbaceous plants, including agricultural crops, is difficult because their host plants vary in abundance and distribution from year to year. Tree-living aphids, in addition to being very host-specific, live in a habitat that is both spatially and temporally relatively stable. Therefore, it is not surprising that most long-term population studies on aphids have been on such species (Dixon 1963, 1966, 1969, 1970, 1971, 1975, 1979, 1990; Dixon and Barlow 1979; Dixon and Mercer 1983; Dixon et al. 1993b; 1996; Barlow and Dixon 1980; Chambers et al. 1985; Wellings et al. 1985). However, some of the theoretical results obtained from these studies are quite general and can be applied to other aphid species.

Within a year, aphid dynamics are very complicated and, in looking for the mechanism of regulation, this needs to be taken into consideration. An initial dramatic increase in population size in spring is typically followed by a steep decline in abundance during summer and sometimes a further increase in autumn (Fig. 1.1). During spring and summer all the generations are parthenogenetic and short lived (1–4 weeks). In autumn, sexual forms develop, mate and give rise to the overwintering eggs from which fundatrices, the first parthenogenetic generation, hatch the following spring. The parthenogenetic generations overlap in time and the environmental conditions are rapidly changing. Therefore, an individual throughout its life, as well as individuals born at different, but close instants in time, can experience quite different conditions, which results in aphids evolving different and varying reproductive strategies.

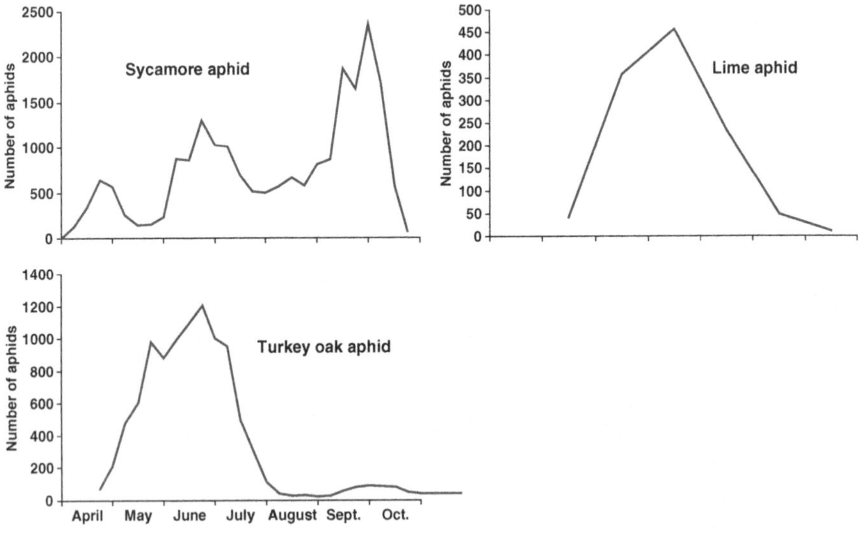

Fig. 1.1 Within season population dynamics of three tree-dwelling aphid species. Values based on direct counts of the number of individuals per 80 leaves, during the years 1960–1974 (Glasgow, UK, sycamore aphid), 1965–1972 (Glasgow, UK, lime aphid) and 1975–1995 (Norwich, UK, Turkey oak aphid). Data collected by A.F.G. Dixon and his students

The within-year dynamics of aphids are largely determined by seasonal changes in host quality. Aphids do best when amino acids are actively translocated in the phloem. In spring, the leaves grow and import amino acids via the phloem; in summer leaves are mature and export mainly sugars. In autumn, the leaves senesce and export amino acids and other nutrients. Thus on trees the leaves are most suitable for aphids in spring and autumn. The differences in within-year population dynamics of aphids are due to differences in the effect these seasonal fluctuations in host plant quality have on the per capita rate of increase and intraspecific competition in each species. This annual cycle, consisting of two short periods when the host plant is very favourable and a long intervening period when it is less favourable, is well documented for tree dwelling aphids. This has greatly facilitated the modelling of their population dynamics. In general the aphid carrying capacity of annual crop plants tends to increase with the season until the plants mature after which it tends to decrease very rapidly. Thus, the aphid carrying capacity of trees tends to be high in spring and autumn and low in summer, whereas that particularly of short-season annual crops tends to be low early in a year, peaking mid year and then declining.

A lot is known about the biology of the parthenogenetic generations of aphids, in particular the optimum behaviour for maximising the instantaneous population growth rate, r_{m}, under various environmental conditions (Kindlmann and Dixon 1989, 1992; Kindlmann et al. 1992) and the optimal strategies for migration (Dixon et al. 1993a). An individual-based model (Kindlmann and Dixon 1996), which incorporates the biology of tree-dwelling aphids, simulated most of the observed features of the population dynamics. It provided a theoretical background for the commonly observed phenomenon that the larger the numbers are at the beginning of a season, the larger and earlier the peak. Migration was shown to be the most important factor determining the summer decline in abundance, while changes in aphid size and food quality account for why the autumnal increase is less steep than in spring. Finally, the model suggests the possibility of a "see-saw effect" (a negative correlation between spring and autumn peak numbers) in some cases, a phenomenon observed in census data (Dixon 1970, 1971).

The regularity of the population fluctuations of the sycamore aphid from year to year: very regular 2-year cycles, as indicated by suction trap catches, has proved very attractive to modellers, who have applied time series analysis to the data (e.g., Turchin 1990; Turchin and Taylor 1992). The conspicuous cyclicity observed in yearly totals of the number of some species of aphid on trees, however, is mainly due to the cyclicity in the peak numbers in spring, which are closely correlated with the yearly totals. It is driven by the inverse relationship between the size of the spring peak and the autumnal rate of increase, the "see-saw effect" (Kindlmann and Dixon 1992). This effect is present in some (Dixon 1971), but not present or very weak in other (Dixon and Kindlmann 1998) empirical data. In *Drepanosiphum platanoidis* (sycamore aphid), where the total numbers on the host tree are relatively constant from year to year, there is a within year see-saw in abundance of aphids in spring and autumn. As most of the aphids that migrate over long distances, rather than between trees, do so in autumn, the result is the 2-year cycles observed in the suction trap catches (Dixon and Kindlmann 1998). Time series and correlation analyses reveal that the spring and late autumn dynamics are often predictable, but not those observed in summer, as the size of the spring peak is not transferred into summer numbers of aphids (Kindlmann and Dixon 1992).

It is argued that aphid population density is regulated by density-dependent processes acting within years, which is reflected in the year-to-year changes in overall abundance (Sequeira and Dixon 1997). Some results suggest a curvilinear density dependence, with strong density-dependent regulation at low densities, and weak at high densities (Jarošík and Dixon 1999).

1.2.2 Biology of Natural Enemies Relevant to Aphid Population Dynamics

Aphidophagous predators, like ladybirds, hoverflies and lacewings commonly occur in agricultural crops, on herbaceous plants and trees. As many feed on aphid pests, their efficiency in controlling them is a widely discussed and controversial issue. Their aphid prey live in colonies, which are characterized by an initial rapid increase followed by an equally rapid decline in abundance resulting in extinction of the colony. The decline is not caused by aphid predators or parasites, even if they contribute to it. Instead, aphids cause the decline: they strongly react to their own density by switching to producing migrants, which disperse and search for another, more suitable host. Thus, when aphid density is high, most of the newborn leave immediately after they mature.

Aphid colonies are characterised by rapid increases and declines in abundance (Dixon 1998), which are not synchronized in time, as the aphids feed on different host plants with different phenologies (Galecka 1966, 1977). On a large spatial scale, at any instant, populations of aphids exist as patches of prey, associated with patches of good host plant quality (Kareiva 1990). That is, aphid predators exploit patches of prey that vary greatly in quality both spatially and temporally and therefore have evolved suitable strategies for effective exploiting this resource.

The adult aphid predator is winged, can easily move between patches of prey, and therefore can find suitable patches of prey. Its immature stages are confined to one patch and if this contains few prey items, the larvae starve and eat each other. Mortality of immature stages due to starvation, cannibalism or intraguild predation is enormous: 98–99% (Osawa 1993; Hironori and Katsuhiro 1997) and is mainly a consequence of low prey numbers that can occur at any time during larval development. Thus egg and larval cannibalism is adaptive, as by eating conspecifics larvae of predators increase their likelihood of survival (Agarwala and Dixon 1992, 1993).

From an evolutionary perspective, both predator and prey species strive to maximize their own reproductive potentials or, more strictly, their genetic fitness. However, whilst prey can exist perfectly well in the absence of predators, predators *require* prey. Therefore, it is in the predator's interest to practice *conservation*. The optimal strategy of the predator then involves the counteracting pressures to maximize its own reproduction and survival, usually through the use of efficient hunting tactics, and yet conserve enough prey for its offspring (Berryman and Kindlmann 2008).

This is nicely exemplified by the case of long-lived insect predators feeding on short-lived prey. As most of these predators suffer an enormous egg and larval mortality due to cannibalism and intraguild predation, selection acts mainly on optimizing their oviposition strategies in terms of maximizing the likelihood that the offspring will survive until reproductive age. The oviposition strategy of a predator with a long larval developmental time relative to that of its prey depends on a longer projection of the future prey abundance in the patch, and therefore the possibility of experiencing more bottlenecks or a higher probability of a bottleneck in prey abundance than a predator with a short developmental time, and consequently must be more conservative in terms of preserving their prey (the "GTR hypothesis"). This "GTR hypothesis" seems to hold more generally and those interested in more details are referred to Kindlmann and Dixon (1999, 2001) and references therein.

Because of the immense egg and larval mortality, selection acts mainly on optimum oviposition strategies – those that insure the maximum likelihood of survival of the offspring – rather than maximization of the food eaten by the predator per unit time, as is assumed to be the case in most optimum foraging theories (Stephens and Krebs 1986). The optimum oviposition strategy of the adult is therefore determined mainly by expectations of future bottlenecks in prey abundance, as these will affect survival of its offspring, and not by the present amount of prey in the patch, as the adult is not limited by the amount of food in the patch as it can find another colony, if needed.

Aphid predators are a good example of the GTR hypothesis, as their developmental time often spans several aphid generations, during which the aphid numbers vary dramatically. Laying eggs in the presence of conspecific larvae is strongly selected against in these predators, because it results in these eggs being eaten by older conspecific larvae. In addition, laying eggs late in the development of an ephemeral patch of prey is maladaptive, as there is insufficient time for the larvae to complete their development. Thus, eggs laid by predators late in the existence of a patch of prey are at a disadvantage, as they are highly likely to be eaten by larvae of predators that hatch from the first eggs to be laid.

The GTR hypothesis is strongly supported by recent empirical results: Mills (2006) and Mills and Latham (2009) show that generation time ratio is an important life history trait that could substantially improve the impact of natural enemies in biological control. They conclude that a small generation time ratio (coupled with a broad window of host attack) can facilitate the suppression of pest abundance by parasitoids, and is positively associated with success in biological control.

Empirical data also indicate that several different species of aphid predators have evolved mechanisms that enable them to oviposit preferentially in patches of prey that are in an early stage of development and avoid those that are already being attacked by larvae (Hemptinne et al. 1992, 1993, 2001). Females of these species strongly react to the larval tracks of their own species or of other aphid predators by immediately ceasing oviposition and flying away from the aphid colony. This response strongly reduces the number of eggs laid per patch and in combination with density dependent cannibalism their effectiveness in regulating the numbers of their prey – aphids. Thus their optimum oviposition strategy, which maximizes the fitness of the individual, results in conserving their prey (or low impact on its numbers), exactly as stated above. Note that for evolution of this strategy no group selection is needed.

Thus oviposition commonly occurs only during a short "egg window", early in the existence of each patch of prey (Hemptinne et al. 1992). When predators are abundant and suitable patches of prey are rare, however, many eggs may nevertheless be laid in a patch during the "egg window". In such circumstances, strong density-dependent cannibalism (Mills 1982) greatly reduces the abundance of the predators relative to that of their prey. Therefore these predators have little impact on aphid population dynamics (Dixon 1992; Kindlmann and Dixon 1993, 1999; Dixon et al. 1995). However, they may nevertheless have short-term impact on local populations, valuable to farmers. The results of an international study of *M. persicae* populations on potato, carried out by 16 workers over 2 years in 10 countries (Mackauer and Way 1976), indicate that the aphid population increased regardless of predator presence, and the latter only affected reductions at times when the potential rate of increase of the aphids was low.

Hymenopterous parasitoids can mature on one aphid and would appear to be potentially more likely to regulate aphid abundance. However, their effectiveness is often reduced by: (1) their longer developmental time relative to their host, (2) the action of hyperparasitoids which, in many cases, are less specific than the primary parasitoids and (3) their vulnerability to attacks from aphid predators (Dixon and Russel 1972; Hamilton 1973, 1974; Holler et al. 1993; Mackauer and Völkl 1993). In addition, because of the risk of hyperparasitism, primary parasitoids are likely to cease ovipositing in a patch where many aphids are already parasitized, as high levels of primary parasitism make the patch attractive to hyperparasitoids. By continuing to oviposit in patches of aphids already attacked by conspecifics these natural enemies may reduce their potential fitness (Ayal and Green 1993; Kindlmann and Dixon 1993).

In the initial phase of aphid population increase on annual arable crops such as spring-sown cereals, there is often a slight dip or plateau followed by sudden acceleration. This is attributed to the activity of polyphagous predators (mainly carabid beetles, spiders and earwigs), and referred to as the "natural enemy ravine" by Southwood and Comins (1976). They suggested that the outcome of a spring invasion of aphids is often determined by the balance between the number of invaders and the size of the autochthonous population of polyphagous predators. Carter and Dixon (1981) offered an alternative explanation: the lack of population growth in the initial phase of the population dynamics was attributed to the intermittent nature of aphid immigration, which is amplified by the pre-reproductive period of the offspring of the immigrant aphids. However, it is more likely that the ravine in population dynamics is a consequence of not being able to detect population increase at low population density using small sample sizes (Jarošík et al. 2003). Small sample sizes were used in the studies cited by Southwood and Comins (1976) as evidence for a natural enemy ravine. In the study of Smith and Hagen (1959) it was 200 alfalfa stems. In that of van Emden (1965) it was 90 mustard plants. Wratten (1975) used 30 stems of wheat. The study of Carter and Dixon (1981), in which an alternative explanation for the ravine was proposed, was also based on small sample sizes, with the maximum sample size of 600 tillers of winter wheat. Honěk and Jarošík (2000) and Honěk et al. (2003) also found no evidence that polyphagous predators affect cereal aphid population dynamics in the field. In the habitat they studied, carabid beetles were the dominant guild of polyphagous predators. However, these carabids are mainly seed predators (Honěk et al. 2003), and their activity was only loosely correlated with aphid density (Honěk and Jarošík 2000). In addition, aphids have a low nutritional value and are not a preferred food of carabids (Bilde and Toft 1999). However, in many crops other than cereals, there is a clear mid-season trough in aphid density between an early and a late peak similar to that which occurs on trees.

1.2.3 Assessment of the Efficiency of Natural Enemies to Suppress the Abundance of Their Prey

Exclusion techniques, such as cages, are the most frequently used means of evaluating the efficiency of natural enemies to suppress the abundance of their prey (Luck et al. 1988). The growth rates and peak densities of aphid populations within cages that exclude natural enemies are usually larger than those in uncaged populations (e.g., Chambers et al. 1983; Elliott and Kieckhefer 2000; Michels et al. 2001; Basky 2003; Cardinale et al. 2003; Schmidt et al. 2003). However, cages change the microenvironment (Hand and Keaster 1967), especially temperature, which is thought to be important in determining the outcome of predator-prey interactions (Frazer and Gilbert 1976; Frazer et al. 1981). This by itself makes the results of cage experiments suspect.

Attempts to avoid the change in the microenvironment by using cages with a large (8 mm) mesh size (Schmidt et al. 2003) do not reduce predator densities within cages and are therefore completely useless for measuring the effect of predators on aphid populations as there are the same numbers of predators inside cages as in un-caged plots (Kindlmann 2010).

Even more importantly, cages prevent aphids from emigrating, which is their usual response to high density (Dixon 1998, 2005). Gardiner et al. (2009) show that after 14 days of caging there are an average of 20.7 ± 1.4 alates per plant within exclusion cages but only 1.8 ± 0.1 alates per plant in un-caged plots. Interestingly, when only polythene enclosures, 60 cm high, buried to a depth of 30 cm, and not cages, were used (Holland et al. 1996), which do not affect the microenvironment of the manipulated plots, allow aphids to emigrate, but exclude ground predators, there was no difference in the number of grain aphids in control plots and those where the number of ground predators were reduced.

Bearing in mind the above, it is therefore really surprising that despite this crystal clear evidence that cage exclusion experiments cannot be used for assessment of predator effectiveness in reducing aphid population growth, they are still used for this purpose and such flawed results are presented as proof of their effectiveness. Other, objective methods of assessing the effectiveness of natural enemies in reducing the abundance of their prey, like removing the predators (Kindlmann et al. 2005) or direct observations like those in Costamagna and Landis (2007), are rare – most likely because they are much more time consuming.

1.3 Theory of Aphid Population Dynamics

1.3.1 Features of Aphid Population Dynamics that Should Be Incorporated in Models

If it is accepted that natural enemies do not regulate aphid populations, the modelling process is greatly simplified. The important features of any model are:

- Each year aphids show an initial dramatic increase in population size.
- This increase is typically followed by a steep decline in abundance.
- Sometimes there is a further increase in abundance.
- Migration is the most important factor determining the decline in abundance.
- Within season aphid dynamics often show a "see-saw effect" – a negative correlation between initial and final peak numbers.
- The greater the initial aphid numbers, the larger and earlier the peak.
- Very regular 2-year cycles are characteristic of aphid between-year population dynamics.
- Aphid population density is regulated by density-dependent processes acting within years, which can be potentially strong at low densities.
- Long term aphid dynamics appear to be little affected by the activity of insect natural enemies.

Here we present a recent metapopulation model of the dynamics of aphids and their predators that satisfies all the above assumptions, which was published by Houdková and Kindlmann (2006), extended by Houdková and Kindlmann (2010) and possibly represents the state-of-the-art in modelling population dynamics of aphids and their predators.

1.3.2 The Metapopulation Model

We consider a fixed number of patches, *p*. The patch may represent a single shoot, one plant, or a patch of these – depending on the mobility of the animals considered. This model has three components:

1. "Egg-Window Dynamics" – the period, when both prey and predators arrive and settle on the plants during spring.
2. "Within-Season Dynamics" – this component follows the previous one and simulates the system dynamics after the initial period of immigration and oviposition has ceased. During this phase, any additional immigration is considered as being small compared to intrinsic system dynamics and therefore neglected.
3. "Between-Season Dynamics" – this component is an iteration of the previous two components and mimics the system behaviour during the course of many years.

1.3.2.1 Egg-Window Dynamics

This component of the model simulates the growth of aphid colonies and the foraging and ovipositing behaviour of predators. We assume that individual prey arrive in the patches only at the beginning of a season and do not migrate between plants. The summer aphid populations are established by parthenogenetic females that emigrate from a winter host-plant, so there is no relationship between numbers of aphids this year and last year on one particular plant. The fundatrices land on plants at random and start to reproduce with a constant growth rate, R. The target plant is chosen from a uniform distribution ($U(1, p)$).

The predators are introduced into the system with an initial amount of energy chosen from a normal distribution with $Emean = 20$, and standard deviation $Esd = 2$. In every step (one flight – we assume several flights per day) all predators are distributed among the plants. The target plant is chosen randomly from the uniform distribution, so it is possible to stay on the same plant and meet other predators. The cost of flight is subtracted from the energy amount of each predator. If there are any aphids on the plant, the predator feeds on them, but does not eat more than a *lunch* (estimated variable) of aphids. Every eaten aphid represents one energy unit, which is added to the predator's energy reserve. If the predator reaches a *min* energy level (*min* – optional variable) it can lay a *batch* of eggs (e.g., 20, estimated variable). A necessary condition for laying eggs is the presence of aphids on the plant and the absence of conspecific larvae, which hatch 4 days after oviposition (*hatch,* estimated variable). The energy used for oviposition is subtracted. The number of aphids eaten is subtracted from the colony. The plants without aphids or with either eggs or larvae are deemed unsuitable for oviposition. Once the proportion of unsuitable plants exceeds a critical value (1–*swPrey*, optional variable), the egg-window closes and the predators leave the system. The maximum length of the egg-window is 15 days. Any eggs that may be laid later and any resultant larvae are unlikely to survive, for the reasons cited above, and are therefore not included in this simulation. The final numbers of prey and predators at the end of the egg-window are used in the within-season dynamics model.

1.3.2.2 Within-Season Dynamics

For simulation of the within-season dynamics we use the Kindlmann and Dixon (1993) model that consists of the following set of differential equations:

$$\frac{dh}{dt} = ax, \qquad h(0) = 0, \qquad \text{(1a) changes in the cumulative density of prey}$$

$$\frac{dx}{dt} = (r - h)x - \frac{vexy}{b + ex + y}, \quad x(0) = x_0, \qquad \text{(1b) changes in prey density}$$

$$\frac{dy}{dt} = -\frac{vy^2}{b + ex + y}, \qquad y(0) = y_0, \qquad \text{(1c) decrease in predator density due to cannibalism}$$

where $h(t)$ is cumulative density of the prey at time t; $x(t)$ is density of prey at time t; a is a scaling constant relating prey cumulative density to its own dynamics; r is maximum potential growth rate of the prey; $y(t)$ is density of predator at time t; v is predator voracity; b is a parameter of the functional response of the predator; e is predator's preference for prey; T is time when predator matures; it coincides with the duration of a patch of prey, yielding final values $x(T) = x_{aut}$ and $y(T) = y_{aut}$ of the season.

This model is based on the following biological observations:

1. Insect herbivores, especially aphids, frequently first increase and then decline in abundance, even in the absence of natural enemies (Dixon 2000). Such declines are often caused by emigration from patches when the prey disperses to find new vacant patches. The prey individuals respond negatively to their cumulative density (Kindlmann et al. 2004). Thus in this model, the regulatory term for prey, when alone, is its cumulative density, h, instead of some function of its instantaneous density. In contrast to the logistic or exponential growth models, this model allows prey to decline in abundance with increasing time even in the absence of natural enemies (Kindlmann et al. 2004).
2. Predators born in a patch rarely reproduce within the same patch (Dixon 2000), but after completing their development leave and reproduce elsewhere. Therefore we assume that (1) the initial density of the predator in a patch is defined by the number of eggs laid there by adults that developed in other patches of prey, arrived to this patch, and reproduced there during the "egg window", and (2) changes over time in the number of predators within a patch are due to larval cannibalism and not reproduction.
3. We assume that the predator is cannibalistic but has a preference, e, for eating prey, as opposed to conspecifics. If they prefer prey, then $e > 1$, but e may also be less than one, as, e.g., in case when the larvae of a predator prefer to eat conspecific eggs, which cannot defend themselves. We have used $e = 1$, the predator shows no preference for either prey or conspecifics (the "meet and eat" hypothesis, Kindlmann and Dixon 1993).
4. The within-season simulation ends with the autumnal host-alternation. As only the alate individuals can reach a winter host-plant and reproduce there, the number of prey next year is derived from the number of alates produced during

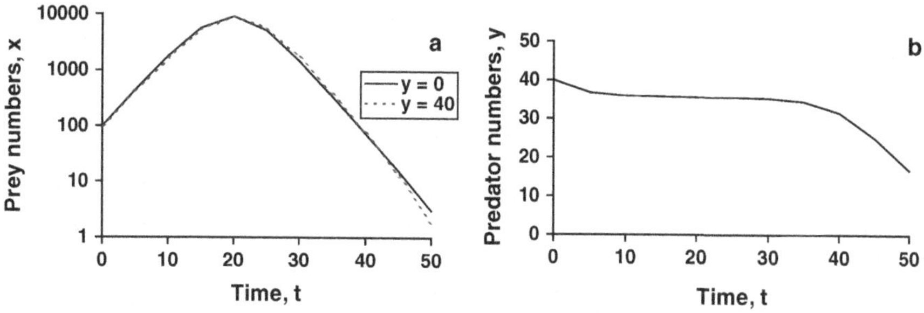

Fig. 1.2 Trends in time in prey (**a**) and predator (**b**) abundance predicted by the model when a = 0.000005, r = 0.3, v = 1, b = 0, p = 1, x_0 = 100, y_0 = 0 and y_0 = 40. In (**a**) prey density in the absence of predators and the presence of 40 predators (see inset) is also presented

the decline in the abundance. All eggs/individuals, which survive winter, will appear in the system in the next season.

5. We consider only female individuals of the predator – supposing a 1:1 sex ratio this does not affect the simulation outcome. The females conform to the optimal ovipositing behaviour outlined above. Larvae do not leave a patch of prey. The length of life of a female is determined by its success in foraging and its ovipositing activity. Once the individual spends all its energy supply, it dies (i.e. leaves the system). The number of predators next year is calculated from the autumn numbers by multiplying by winter mortality and only a half of that is used as the number of females.

A typical trend in numbers in a patch predicted by model (1) is shown in Fig. 1.2. There is only one peak in the abundance of prey and the individuals respond negatively to their cumulative density resulting in the bottleneck in their abundance. There is no predator reproduction in the patch; therefore, predator numbers monotonously decline. As a consequence, if prey abundance (x) increases at the beginning (i.e., if y_0 is sufficiently small, so that $\lim_{t\to 0+} \frac{dx}{dt} > 0$) then as time proceeds, the dynamics of the prey is less and less influenced by the declining numbers of the predator. Because of the way the diet of the predator is defined (the terms containing v in (1b) and (1c)), the decline in predator numbers is more pronounced when there are few prey individuals relative to predator individuals – that is, when the ratio x/y is small at the beginning and when prey numbers have passed their peak and become small again due to the negative effect of cumulative density. Within a season, the predators have almost no influence on the prey dynamics in this system (Houdková and Kindlmann 2006). Not surprisingly, the number of predators that survive is positively influenced by the initial number of prey and negatively influenced by the initial number of predators (Houdková and Kindlmann 2006).

The predicted within-season trends in abundance (Houdková and Kindlmann 2006, 2010) closely match those observed in nature in aphids (Dixon et al. 1996; Kindlmann and Dixon 1996, 1997; Dixon and Kindlmann 1998; Kindlmann et al.

2005) and ladybird beetles (Osawa 1993; Hironori and Katsuhiro 1997; Yasuda and Ohnuma 1999; Kindlmann et al. 2005).

1.3.2.3 Between-Season Dynamics

The total number of winged aphids and adult predators produced in all patches, reduced by a winter mortality factor (assumed to be constant), are assumed to create the next year metapopulation of aphids and predators and are used as input data for the egg-window phase next spring. Mathematically, the autumn numbers of prey and predators are multiplied by their probabilities of survival over winter $\boldsymbol{P}_x$ and $\boldsymbol{P}_y$, respectively.

1.3.2.4 Model Predictions

The predictions of this model are rather complicated and summarized in Houdková and Kindlmann (2006, 2010). Cycles similar to those observed in many studies are predicted. The main result is, however, that natural enemies are unlikely to affect aphid population dynamics during the season, but may have an effect very early in the season, when aphid colonies are still small and predators might be able to reduce the numbers of these colonies. Empirical verification of this is still very weak, however. The only long-term studies on parallel oscillations of aphid and coccinellid abundance are those of Carter et al. (1982) and Honěk and Martinková (2005). They show that the abundance of ladybirds recorded at hibernation sites is correlated with the abundance of cereal aphids infesting winter wheat the previous summer. Other studies on ladybirds report similar annual fluctuations in abundance (Elliott and Kieckhefer 1990; Kieckhefer and Elliott 1990). Further empirical results of this kind are needed, however.

1.4 Predictions

Despite the enormous effort that has been devoted to modelling aphid dynamics, there is no simple and reliable long-term model suitable for predicting aphid abundance. The existing models can sometimes make short-term predictions, which are of little practical value. Long-term, between-year predictions, however, require a metapopulation approach, because aphid migration, which occurs at least twice a year, scrambles the results of population dynamics on individual plants. It is also clear that heterogeneities in the landscape and how they change in time also need to be incorporated into the models as they can have a marked effect on the local dynamics of aphids (Dedryver et al. 2009). As indicated in Section 1.3, such models are complicated and currently it is not clear whether they can be simplified and yet still result in meaningful predictions. However, it is clear that if the model is complicated then farmers will not use it. Verification of such models is an even bigger problem as long-term, large-scale data on population dynamics of aphids and their enemies are scarce.

The situation becomes even more complicated, when the possible effect of global change is incorporated in the models. The expectation is that this will result in changes in the distribution of aphids, which have already occurred for some species. Also, as increases in temperature speed up processes in all the trophic levels it is also likely that the outcome of the dynamics will be different. In the absence of information on the relative effects of such changes on species in each of the trophic levels, however it is not possible to predict the outcome using mathematical models of population dynamics. Hopefully, further empirical data will reverse this rather pessimistic conclusion.

Acknowledgements The work was supported by the grant No. LC 06073 of the MSMT.

References

Agarwala BK, Dixon AFG (1992) Laboratory study of cannibalism and interspecific predation in ladybirds. Ecol Entomol 17:303–309

Agarwala BK, Dixon AFG (1993) Kin recognition: egg and larval cannibalism in *Adalia bipunctata* (Coleoptera: Coccinellidae). Eur J Entomol 90:45–50

Ayal Y, Green RF (1993) Optimal egg distribution among host patches for parasitoids subject to attack by hyperparasitoids. Am Nat 141:120–138

Barlow ND, Dixon AFG (1980) Simulation of lime aphid population dynamics. Pudoc, Wageningen

Basky Z (2003) Predators and parasitoids on different cereal aphid species under caged and no caged conditions in Hungary. In: Soares AO, Ventura MA, Garcia V, Hemptinne J-L (eds) Proceedings of the 8th international symposium on ecology of aphidophaga: biology, ecology and behaviour of aphidophagous insects. Arquipélago – Life and Marine Science, Supplement 5, Ponta Delgada, Azores, Portugal

Berryman AA, Kindlmann P (2008) Population systems: a general introduction. Springer, Dordrecht

Bilde T, Toft S (1999) Prey consumption and fecundity of the carabid beetle *Calathus melanocephalus* on diets of three cereal aphids: high consumption rates of low quality prey. Pedobiologia 43:422–429

Cardinale BJ, Harvey CT, Gross K, Ives AR (2003) Biodiversity and biocontrol: emergent impacts of a multi-enemy assemblage on pest suppression and crop yield in an agroecosystems. Ecol Lett 6:857–865

Carter N (1985) Simulation modelling of the population dynamics of cereal aphids. Biosystems 1:111–119

Carter N, Dixon AFG (1981) The "natural enemy ravine" in cereal aphid population dynamics. J Anim Ecol 50:605–611

Carter N, Dixon AFG, Rabbinge R (1982) Cereal aphid populations: biology, simulation and predation. Centre for Agricultural Publishing, Wageningen

Chambers RJ, Sunderland KD, Wyatt IJ, Vickerman GP (1983) The effects of predator exclusion and caging on cereal aphids in winter wheat. J Appl Ecol 20:209–224

Chambers RJ, Wellings PW, Dixon AFG (1985) Sycamore aphid numbers and population density 11. Some processes. J Anim Ecol 54:425–442

Costamagna AC, Landis DA (2007) Quantifying predation on soybean aphid through direct field observations. Biol Control 42:16–24

DeLoach CJ (1974) Rate of increase of populations of cabbage, green peach, and turnip aphids at constant temperatures. Ann Entomol Soc Am 67:332–340

Dedryver CA, Fiévet V, Plantegnest M, ad Vialatte A (2009) An overview of *Sitobion avenae* population functioning at three spatial scales in France. Redia 42:159–162

Dixon AFG (1963) Reproductive activity of the sycamore aphid, *Drepanosiphum platanoides* (Schr.) (Hemiptera. Aphididae). J Anim Ecol 32:33–48
Dixon AFG (1966) The effect of population density and nutritive status of the host on the summer reproductive activity of the sycamore aphid, *Drepanosiphum platanoides* (Schr.). J Anim Ecol 35:105–112
Dixon AFG (1969) Population dynamics of the sycamore aphid *Drepanosiphum platanoides* (Schr.) (Hemiptera: Aphididae): Migratory and trivial flight activity. J Anim Ecol 38:585–606
Dixon AFG (1970) Stabilisation of aphid populations by an aphid induced plant factor. Nature (London) 227:1368–1369
Dixon AFG (1971) The role of intra-specific mechanisms and predation in regulating the numbers of the lime aphid, *Eucallipterus tiliae* L. Oecologia 8:179–193
Dixon AFG (1975) Effect of population density and food quality on autumnal reproductive activity in the sycamore aphid, *Drepanosiphum platanoides* (Schr). J Anim Ecol 44:297–304
Dixon AFG (1979) Sycamore aphid numbers: the role of weather, host and aphid. In: Anderson RM, BD Turner, LR Taylor (eds) Population dynamics. Blackwell, Oxford, pp 105–121
Dixon AFG (1990) Population dynamics and abundance of deciduous tree-dwelling aphids. In: Hunter M, N Kidd, SR Leather, AD Watt (eds) Population dynamics of forest insects. Intercept, Andover, pp 11–23
Dixon AFG (1992) Constraints on the rate of parthenogenetic reproduction and pest status of aphids. Invertebr Reprod Dev 22:159–163
Dixon AFG (1998) Aphid ecology: an optimization approach, 2nd edn. Chapman and Hall, London
Dixon AFG (2000) Insect predator-prey dynamics: ladybird beetles and biological control. Cambridge University Press, Cambridge
Dixon AFG (2005) Insect herbivore-host dynamics: shrub-dwelling aphids. Cambridge University Press, Cambridge
Dixon AFG, Barlow ND (1979) Population regulation in the lime aphid. Zool J Linn Soc 67: 225–237
Dixon AFG, Hemptinne J-L, Kindlmann P (1995) The ladybird fantasy – prospects and limits to their use in the biological control of aphids. Züchtungsforschung 1:395–397
Dixon AFG, Horth S, Kindlmann P (1993a) Migration in insects: costs and strategies. J Anim Ecol 62:182–190
Dixon AFG, Kindlmann P (1998) Population dynamics of aphids. In: Dempster PJ, IFG McLean (eds) Insect populations. Kluwer, Dordrecht, pp 207–230
Dixon AFG, Kindlmann P, Sequeira R (1996) Population regulation in aphids. In: Floyd RB, AW Sheppard, PJ De Barro (eds) Frontiers of population ecology. CSIRO Publishing, Melbourne, pp 103–114
Dixon AFG, Mercer DR (1983) Flight behaviour in the sycamore aphid: factors affecting take-off. Entomol Exp Appl 33:43–49
Dixon AFG, Russel RJ (1972) The effectiveness of *Anthocoris nemorum* and *A. confusus* (Hemiptera: Anthocoridae) as predators of the sycamore aphid, *Drepanosiphum platanoides* II. Searching behaviour and the incidence of predation in the field. Entomol Exp Appl 15:35–50
Dixon AFG, Wellings PW, Carter C, Nichols JFA (1993b) The role of food quality and competition in shaping the seasonal cycle in the reproductive activity of the sycamore aphid. Oecologia 95:89–92
Elliott NC, Kieckhefer RW (1990) Dynamics of aphidophagous coccinellid assemblages in small grain fields in eastern South Dakota. Environ Entomol 19:1320–1329
Elliott NC, Kieckhefer RW (2000) Response by coccinellids to spatial variation in cereal aphid density. Popul Ecol 42:81–90
van Emden HF (1965) The effect of uncultivated land on the distribution of cabbage aphid (*Brevicorine brassicae*) on an adjacent crop. J Appl Ecol 2:171–196
Frazer BD, Gilbert N (1976) Coccinellids and aphids: a quantitative study of the impact of adult ladybirds (Coleoptera: Coccinellidae) preying on field populations of pea aphids (Homoptera: Aphididae). J Entomol Soc B C 73:33–56

Frazer BD, Gilbert N, Ives PM, Raworth DA (1981) Predator reproduction and the overall predator-prey relationship. Can Entomol 113:1015–1024

Galecka B (1966) The role of predators in the reduction of two species of potato aphids, *Aphis nasturtii* Kalt. and *A. frangulae* Kalt. Ekologia Polska 14:245–274

Galecka B (1977) Effect of aphid feeding on the water uptake by plants and on their biomass. Ekologia Polska 25:531–537

Gardiner MM, Landis DA, Gratton C, Difonzo CD, O'Neal M, Chacon JM, Wayo MT, Schmidt NP, Mueller EE, Heimpel GE (2009) Landscape diversity enhances biological control of an introduced crop pest in the north-central USA. Ecol Appl 19:143–154

Gilbert N, Hughes RD (1971) A model of an aphid population – three adventures. J Anim Ecol 40:525–534

Gonzalez-Andujara JL, Garcia-de Cecab JL, Fereresc A (1993) Cereal aphids expert system (CAES): identification and decision making. Comput Electron Agric 8:293–300

Gosselke U, Triltsch H, Roßberg D, Freier B (2001) GETLAUS01 – the latest version of a model for simulating aphid population dynamics in dependence on antagonists in wheat. Ecol Model 145:143–157

Gutierrez AP, Baumgaertner JU (1984a) Multitrophic level models of predator-prey energetics I. Age specific energetics models – pea aphid *Acyrthosiphum pisum* (Harris) (Homoptera: Aphididae) as an example. Can Entomol 116:924–932

Gutierrez AP, Baumgaertner JU (1984b) Multitrophic level models of predator-prey energetics II. A realistic model of plant-hrebivore-parasitoid-predator interactions. Can Entomol 116: 933–949

Gutierrez AP, Baumgaertner JU, Summers GC (1984) Multitrophic level models of predator-prey energetics III. A case study of an alfalfa system. Can Entomol 116:950–963

Gutierrez AP, Baumgärtner JU, Hagen KS (1981) A conceptual-model for growth, development, and reproduction in the ladybird beetle, *Hippodamia convergens* (Coleoptera, Coccinellidae). Can Entomol 113:21–33

Gutierrez AP, Havenstein DE, Nix HA, Moore PA (1974) Ecology of *Aphis craccivora* Koch and subterranean clover stunt virus in southeast Australia. 2. Model of cowpea aphid populations in temperate pastures. J Appl Ecol 11:1–20

Gutierrez AP, Summers CG, Baumgaertner J (1980) The phenology and distribution of aphids in California alfalfa as modified by ladybird beetle predation (Coleoptera, Coccinellidae). Can Entomol 112:489–495

Hamilton PA (1973) The biology of *Aphelinus flavus* (Hym. Aphelinidae), a parasite of the sycamore aphid *Drepanosiphum platanoidis* (Hemipt. Aphididae). Entomophaga 18:449–462

Hamilton PA (1974) The biology of *Monoctonus pseudoplatani*, *Trioxys cirsii* and *Dyscritulus planiceps*, with notes on their effectiveness as parasites of the sycamore aphid, *Drepanosiphum platanoidis*. Ann Soc Entomol Fr 10:821–840

Hand LF, Keaster AJ (1967) The environment of an insect field cage. J Econ Entomol 60:910–915

Harrington R, Hullé M, Plantegenest M (2007) Monitoring and forecasting. In: van Emden HF, R Harrington (eds) Aphids as crop pests. CABI International, London, pp 515–536

Hemptinne J-L, Dixon AFG, Coffin J (1992) Attack strategy of ladybird beetles (Coccinellidae): factors shaping their numerical response. Oecologia 90:238–245

Hemptinne J-L, Dixon AFG, Doucet JL, Petersen JE (1993) Optimal foraging by hoverflies (Diptera: Syrphidae) and ladybirds (Coleoptera: Coccinellidae): mechanisms. Eur J Entomol 90:451–455

Hemptinne J-L, Lognay G, Doumbia M, Dixon AFG (2001) Chemical nature and persistence of the oviposition deterring pheromone in the tracks of the larvae of the two spot ladybird, *Adalia bipunctata* (Coleoptera: Coccinellidae). Chemoecology 11:43–47

Hironori Y, Katsuhiro S (1997) Cannibalism and interspecific predation in two predatory ladybird beetles in relation to prey abundance in the field. Entomophaga 42:153–163

Holland JM, Thomas SR, Hewitt A (1996) Some effects of polyphagous predators on an outbreak of cereal aphid (*Sitobion avenae* F.) and orange wheat blossom midge (*Sitodoplosis mosellana* Gehin). Agric Ecosyst Environ 59:181–190

Holler C, Borgemeister C, Haardt H, Powell W (1993) The relationship between primary parasitoids and hyperparasitoids of cereal aphids: an analysis of field data. J Anim Ecol 62:12–21

Honěk A, Jarošík V (2000) The role of crop density, seed and aphid presence in diversification of field communities of Carabidae (Coleoptera). Eur J Entomol 97:517–525

Honěk A, Martinková Z (2005) Long term changes in abundance of *Coccinella septempunctata* (Coleoptera: Coccinellidae) in the Czech Republic. Eur J Entomol 102:443–448

Honěk A, Martinková J, Jarošík V (2003) Carabids as seed predators. Eur J Entomol 100:531–544

Houdková K, Kindlmann P (2006) Scaling up population dynamic processes in a ladybird-aphid system. Popul Ecol 48:323–332

Houdková K, Kindlmann P (2010) Modelling of aphid-coccinellid interactions on a long-term scale. Submitted

Hughes RD (1963) Population dynamics of the cabbage aphid, *Brevicoryne brassicae* (L). J Anim Ecol 32:393–424

Hughes RD, Gilbert N (1968) A model of an aphid population. J Anim Ecol 37:553–563

Huxley TH (1858) On the agamic reproduction and morphology of Aphis – Part 1. Trans Linn Soc 22:193–219

Jarošík V, Dixon AFG (1999) Population dynamics of a tree-dwelling aphid: regulation and density independent processes. J Anim Ecol 68:726–732

Jarošík V, Honěk A, Dixon AFG (2003) Natural enemy ravine revisited: the importance of sample size for determining population growth. Ecol Entomol 28:85–91

Kareiva P (1990) Population dynamics in spatially complex environments: theory and data. Philos Trans R Soc Lond B 330:175–190

Kieckhefer RW, Elliott NC (1990) A 13-year survey of the aphidophagous Coccinellidae in maize fields in eastern South Dakota. Can Entomol 122:579–581

Kindlmann (2010) Some cage exclusion experiments do not accurately measure the effectiveness of predators in reducing the abundance of their prey. Submitted

Kindlmann P, Dixon AFG (1989) Developmental constraints in the evolution of reproductive strategies: telescoping of generations in parthenogenetic aphids. Funct Ecol 3:531–537

Kindlmann P, Dixon AFG (1992) Optimum body size: effects of food quality and temperature, when reproductive growth rate is restricted. J Evol Biol 5:677–690

Kindlmann P, Dixon AFG (1993) Optimal foraging in ladybird beetles (Coleoptera: Coccinellidae) and its consequences for their use in biological control. Eur J Entomol 90:443–450

Kindlmann P, Dixon AFG (1996) Population dynamics of a tree-dwelling aphid: individuals to populations. Ecol Model 89:23–30

Kindlmann P, Dixon AFG (1997) Patterns in the population dynamics of the Turkey-oak aphid. In: Nieto Nafria JM, AFG Dixon (eds) Aphids in natural and managed ecosystems. Universidad de León, León, pp 219–225

Kindlmann P, Dixon AFG (1999) Generation time ratios – determinants of prey abundance in insect predator-prey interactions. Biol Control 16:133–138

Kindlmann P, Dixon AFG (2001) When and why top-down regulation fails in arthropod predator-prey systems. Basic Appl Ecol 2:333–340

Kindlmann P, Dixon AFG, Gross LJ (1992) The relationship between individual and population growth rates in multicellular organisms. J Theor Biol 157:535–542

Kindlmann P, Arditi R, Dixon AFG (2004) A simple aphid population model. In: Simon J-C, Dedryver C-A, Rispe C, Hulle M (eds) Aphids in a new millenium. INRA Editions, INRA, Paris, pp 325–330

Kindlmann P, Yasuda H, Kajita Y, Dixon AFG (2005) Field test of the effectiveness of ladybirds in controlling aphids. In: Hoddle MS (ed) International symposium on biological control of arthropods, September 12–16, 2005. USDA Forest Service, FHTET–2005–08, Davos, Switzerland, pp 441–447

Knudsen GR, Schotzko DJ (1991) Simulation of Russian wheat aphid movement and population-dynamics on preferred and nonpreferred host plants. Ecol Model 57:117–131

Luck RF, Shepard BM, Kenmore PE (1988) Experimental methods for evaluating arthropod natural enemies. Ann Rev Entomol 33:367–391
Mack TP, Smilowitz Z (1981) The vertical distribution of green peach aphids and its effect on a model quantifying the relationship between green peach aphids and a predator. Am Potato J 58:345–353
Mack TP, Smilowitz Z (1982) Using temperature-mediated functional response models to predict the impact of *Coleomegilla maculata* (DeGeer) adults and 3rd-instar larvae on green peach aphids. Environ Entomol 11:46–52
Mackauer M, Völkl W (1993) Regulation of aphid populations by aphidiid wasps: does parasitoid foraging behaviour or hyperparasitism limit impact? Oecologia 94:339–350
Mackauer M, Way MJ (1976) Myzus persicae Sulz. an aphid of world importance. In: Delucchi VL (ed) Studies in biological control. Cambridge University Press, Cambridge, pp 51–119
Mann BP, Wratten SD, Watt AD (1986) A computer-based advisory system for cereal aphid control. Comput Electron Agric 1:263–270
Michels GJ, Elliott NC, Romero RA, Owings DA, Bible JB (2001) Impact of indigenous coccinellids on Russian wheat aphids and greenbugs (Homoptera: Aphididae) infesting winter wheat in the Texas Panhandle. Southwest Entomol 26:97–114
Mills NJ (1982) Voracity, cannibalism and coccinellid predation. Ann Appl Biol 101:144–148
Mills NJ (2006) Accounting for differential success in the biological control of homopteran and lepidopteran pests. N Z J Ecol 30:61–72
Mills NJ, Latham D (2009) Quantifying the role of predation in the seasonal dynamics of mealy plum aphid populations in California. Redia, 42:153–157
Morgan D (2000) Population dynamics of the bird cherry-oat aphid, *Rhopalosiphum padi* (L.), during the autumn and winter: a modelling approach. Agric For Entomol 2:297–304
Osawa N (1993) Population field studies of the aphidophagous ladybird beetle *Harmonia axyridis* (Coleoptera: Coccinellidae): life tables and key factor analysis. Res Popul Ecol 35:335–348
de Réaumur RP (1737) Mémoires pour servir à l'histoire des insects, vol 111. Imprimerie Royale, Paris, pp 332–350
Ro TH, Long GE (1998) Population dynamics pattern of green peach aphid (Homoptera: Aphididae) and its predator complex in a potato system. Korean J Biol Sci 2:217–222
Ro TH, Long GE (1999) GPA-Phenodynamics, a simulation model for the population dynamics and phenology of green peach aphid in potato: formulation, validation, and analysis. Ecol Model 119:197–209
Schmidt MH, Lauer A, Purtauf T, Thies C, Schaefer M, Tscharntke T (2003) Relative importance of predators and parasitoids for cereal aphid control. Proc R Soc Lond B 270:1905–1909
Scopes NEA (1969) The potential of *Chrysopa carnea* as a biological control agent of *Myzus persicae* on glasshouse chrysanthemums. Ann Appl Biol 64:433–439
Sequeira R, Dixon AFG (1997) Patterns and processes in the population dynamics of tree-dwelling aphids: the importance of time scales. Ecology 78:2603–2610
Skirvin DJ, Perry JN, Harrington R (1997a) A model describing the population dynamics of *Sitobion avenae* and *Coccinella septempunctata*. Ecol Model 96:29–39
Skirvin DJ, Perry JN, Harrington R (1997b) The effect of climate change on an aphid–coccinellid interaction. Glob Change Biol 3:1–11
Smilowitz Z (1984) GPA-CAST: a computerized model for green peach aphid management on potatoes. In: Lashomb JH, R Casagrande (eds) Advances in potato pest management. Hutchinson Ross, Pennsylvania, pp 193–203
Smith RF, Hagen KS (1959) Impact of commercial insecticide treatments. Hilgardia 29:131–154
Southwood TRE, Comins HN (1976) A synoptic population model. J Anim Ecol 45:949–965
Stephens DW, Krebs JR (1986) Foraging theory. Princeton University Press, Princeton, NJ
Stewart L, Dixon AFG (1988) Quantification of the effect of natural enemies on the abundance of aphids: precision and statistics. In: Niemczyk E and AFG Dixon (eds) Ecology and effectiveness of aphidophaga. Academic Publishing, The Hague, pp 187–197
Tamaki G (1973) Spring populations of the green peach aphid on peach trees and the role of natural enemies in their control. Environ Entomol 2:186–191

Tamaki G (1984) Biological control of potato pests. In: Lashomb JH, R Casagrande (eds) Advances in potato pest management. Hutchinson Ross, Pennsylvania, pp 178–192

Tamaki G, Weeks RE (1972) Efficiency of three predators, *Geocoris bullatus*, *Nabis americoferous*, and *Coccinella transversoguttata*, used alone or in combination against three prey species, *Myzus persicae*, *Ceramica picta*, and *Mamestra configurata*, in a greenhouse study. Environ Entomol 1:258–263

Tamaki G, Weeks RE (1973) The impact of predators on populations of green peach aphids on field-grown sugar beets. Environ Entomol 2:345–349

Tamaki G, Weiss MA, Long GE (1980) Impact of high temperatures on the population dynamics of the green peach aphid in field cages. Environ Entomol 9:331–337

Tamaki G, Weiss MA, Long GE (1982) Effective growth units in population dynamics of the green peach aphid (Homoptera: Aphididae). Environ Entomol 11:1134–1136

Taylor LR (1977) Migration and the spatial dynamics of an aphid, *Myzus persicae*. J Anim Ecol 46:411–423

Turchin P (1990) Rarity of density dependence or population regulation with lags? Nature 344:660–663

Turchin P, Taylor AD (1992) Complex dynamics in ecological time series. Ecology 73:289–305

Watt AD (1983) The influence of forecasting on cereal aphid control strategies. Crop Prot 2: 417–429

Watt AD, Vickerman GP, Wratten SD (1984) The effect of the grain aphid, *Sitobion avenae* (F.), on winter wheat in England: an analysis of the economics of control practice and forecasting systems. Crop Prot 3:209–222

Weisser WW (2000) Metapopulation dynamics in an aphid-parasitoid system. Entomol Exp Appl 97:83–92

Wellings PW, Chambers RJ, Dixon AFG, Aikman DP (1985) Sycamore aphid numbers and population density 1. Some patterns. J Anim Ecol 54:411–424

Whalon ME, Smilowitz Z (1979) Temperature-dependent model for predicting field populations of green peach aphid, *Myzus persicae* (Sulzer) (Homoptera: Aphididae). Can Entomol 111: 1025–1032

Wiktelius S, Pettersson J (1985) Simulations of bird cherry–oat aphid population dynamics: a tool for developing strategies for breeding aphid-resistant plants. Agric Ecosyst Environ 14: 159–170

Williams IS, Dixon AFG (2007) Life cycles and polymorphism. In: van Emden HF, R Harrington (eds) Aphids as crop pests. CABI International, London

Winder L, Alexander CJ, Holland JM, Woolley C, Perry JN (2001) Modelling the dynamic spatio-temporal response of predators to transient prey patches in the field. Ecol Lett 4:568–576

Woolhouse MEJ, Harmsen R (1991) Population dynamics of *Aphis pomi*: a transition matrix approach. Ecol Model 55:103–111

Wratten SD (1975) The nature of the effects of aphids *Sitobion avenae* and *Metopolophium dirhodum* on the growth of wheat. Ann Appl Biol 79:27–34

Yasuda H, Ohnuma N (1999) Effect of cannibalism and predation on the larval performance of two ladybirds. Entomol Exp Appl 93:63–67

Chapter 2
Aphids in a Changing World

Olga M.C.C. Ameixa

Abstract When in 1824 the French mathematician Jean Baptiste Joseph Fourier suggested that the Earth's temperature was slowly increasing it was not readily accepted. Some years later Svante Arrhenius (1896) and Guy Callendar (1938) supported this hypothesis and added that the planet's temperature is increasing due to man's activities, in particular the production of CO_2, which has a crucial role in this matter. Several authors question how global warming is going to affect the planet and life forms. This question is addressed here, in particular how it is likely to influence aphids. Will they move to different locations, adapt to the change in conditions in their current habitat or go extinct? An analysis of the literature on the reactions to the sort of changes aphids will be exposed to in global change scenarios revealed that the results are contradictory, indicating positive, negative or no effect. The consensus is that it is extremely difficult to do experiments that simulate future conditions, especially when more than one parameter is considered. Nevertheless, the paleontological data provide a good record of how insects have responded to previous climatic changes. The evidence indicates that aphids are most likely to move to different geographical locations in order to track more suitable conditions, which may be more difficult than in the past because of habitat fragmentation and habitat loss.

Keywords Aphids · Climate change · Adaptation · Behaviour · Host plant · Population dynamics · Extinction · Distribution · Trophic interactions

---

O.M.C.C. Ameixa (✉)
Department of Theoretical Ecology, Institute of Systems Biology and Ecology AS CR, CZ-370 05 České Budějovice, Czech Republic
e-mail: olgameixa@portugalmail.pt

P. Kindlmann et al. (eds.), *Aphid Biodiversity under Environmental Change*,
DOI 10.1007/978-90-481-8601-3_2, © Springer Science+Business Media B.V. 2010

2.1 Introduction

Planet Earth has experienced many significant climatic changes. One of the most important parameters associated with climatic change is the carbon dioxide (CO_2) concentration in the atmosphere. Nevertheless, other greenhouse gases like tropospheric ozone (O_3), methane and water vapour contribute to global warming. The increase in the concentration of these gases in the atmosphere induces an increase in temperature, which influences other climatic parameters. The records of Paleoclimate show that climate has changed abruptly in the past and that changes during the Earth glacial cycles are closely associated with the level of CO_2 in the atmosphere.

Anthropogenic effects on the climate are a relatively recent phenomenon. Although such changes have occurred frequently in the history of planet Earth, the effect due to human activities has increased markedly since the beginning of the Industrial Revolution (1750). Since then, the concentration of tropospheric O_3 has increased by 36%. In the northern hemisphere O_3 concentrations are increasing by 0.5–2.5% per year (Prather and Ehhalt 2001; Prentice 2001). Predictions indicate that in the next century the planet will experience a doubling of atmospheric CO_2 to 700 ppm and a rise in global mean temperatures by 1.4–5.8°C by 2100 (IPCC 2001). Global mean surface temperatures have increased by approximately 0.6°C in the last century, the largest 100-year increase for the last 1,000 years (Houghton et al. 2001).

Biodiversity is being affected by these climatic changes in several ways: life forms can evolve, shift their location or in a worst scenario go extinct. If the current predictions are correct, the rate of environmental changes may outpace the capacities of organisms to adapt to the changes (Flynn et al. 2006).

Future anthropogenic induced changes will influence regional geographic patterns of climate, like temperature and precipitation and these play a crucial role in determining the geographical distribution of species and ecological communities. It is well documented that a gradual change in climate, as well as local or regional climate characteristics, can affect population abundance (Singer and Thomas 1996; Martin 1998), species' distribution (Andrewartha and Birch 1954; Woodward 1987; Root 1988; Davis and Zabinski 1992; Coope 1995; Parmesan 1996), morphology (Hadly 1997), behaviour (Rubenstein 1992) and ultimately also community structure (Pickett and White 1985).

Changes in the world's climate have the potential to greatly alter the relationships between plants and insects, with potential areas of influence ranging from patterns of biodiversity to agricultural productivity (Körner 2000; Theurillat and Guisan 2001; Körner 2003). In such scenarios, insect herbivores, which annually consume 10–15% of global net primary productivity (Crawley 1983), will be both directly and indirectly affected. Phloem-feeding insects are the only species showing a positive response to plants grown in conditions of elevated levels of CO_2, although the response is species-specific (Newman et al. 1999; Whittaker 1999; Holopainen 2002; Williams et al. 2003), with even the same species responding differently on different host plants (Awmack et al. 1997).

Climate change may jeopardize the services provided by ecosystems, such as pollination of crops, pest control by predators, water purification or soil renewal. These changes affect not only the interactions between insects and plants but also natural and agricultural ecosystems (Körner 2000; Theurillat and Guisan 2001).

Some studies indicate that current changes, especially the elevated concentrations of O_3 and CO_2 (IPCC 2001), may affect the performance (e.g., growth and fecundity) and population dynamics of aphids (Holopainen 2002; Percy et al. 2002; Awmack et al. 2004). Aphids are a good model for studying the effects of environmental change since they have short life cycles, with several generations per year, and a high fecundity (Dixon 1985). The life histories of aphids are well studied, but the range of factors that can influence their performance is so great that the conclusions drawn from these studies tend to be contradictory. Bottom-up approaches to determining aphid performance using a plant or even a single detached leaf, for example, indicate that increased concentrations of atmospheric CO_2 might directly affect aphids and their host plant. The results of these studies are extrapolated to ecosystems. But is this the correct way to study phenomena that are of a global nature? Determining the relationship between aphids and their plant-host in changing conditions is challenging, but the results of such studies are contradictory. For this reason, several authors had suggested that every aphid-plant interaction might be unique and unpredictable. Because of this species-specific dependence it is difficult to predict community-level responses. It is certain that plants are affected by changes in temperature and greenhouse gases, which has consequences for the behaviour of phytophagous insects.

The objective of this chapter is to indicate the likely response of aphids to the predicted climate change by reviewing some of the empirical studies, especially the results of long-term studies.

2.2 Empirical Studies on the Behaviour of Aphids

It is a very difficult task to draw a general conclusion about how aphid performance is likely to respond to the predicted global change scenario based on empirical studies. Listed in Table 2.1 are some of the most important empirical and long-term studies. It is difficult to draw any general conclusions about how aphids will respond to future climatic changes from these results. This is because these studies have usually only determined the effect of one or two of the parameters associated with global warming, and the responses of the plants, aphids and natural enemies depend on the "parameter selection". However, in the history of our planet there have been several major climatic changes since aphids evolved, so looking at what happened in the past could throw light on what is likely to happen in the future.

It is known that in the past there were periods of climate instability during which there were dramatic and fast climate changes. Studies of the ratio of the two oxygen molecules with different molecular weights (O_{16} and O_{18} isotopes) in ice cores provide a proxy record of the temperature conditions that existed when the snow was deposited. These ice cores also provide information on the presence

Table 2.1 Published reports of empirical studies on the effects of climate change on aphids (– means that these parameters were not tested in the study)

Species	Plant-host	CO_2	O_3	T	P	D	N	References	Behavioural response
Sitobion avenae	*Triticum* spp.	700 ppm	–	–	–	–	–	Awmack et al. (1996)	Increased fecundity
Aulacorthum solani	*Vicia faba*	350 pm/700 ppm	–	–	–	–	–	Awmack et al. (1997)	16% increase in fecundity
Aulacorthum solani	*Tanacetum vulgare*								10% faster development
Cepegillettea betulaefoliae	*Betula papyrifera*	amb. +200 ppm	1.5 × amb.	1.5 × amb.	–	–	–	Awmack et al. (2004)	Aphid individual performance did not predict population responses – Elevated CO_2 and O_3 are unlikely to affect *C. betulaefoliae* populations in the presence of natural enemies
Acyrtosiphon pisum *Aulacorthum solani*	*Vicia faba*	700 ppm	–	–	–	–	–	Awmack and Harrington (2000)	No effect on aphid density

Table 2.1 (continued)

Species	Plant-host	CO_2	O_3	T	P	D	N	References	Behavioural response
Myzus persicae	*Brassica oleracea*	amb. +200 ppm		–	–	–	–	(Bezemer et al., 1999)	Increased fecundity on *Brassica oleracea*
Myzus persicae	*Senecio vulgaris*								No fecundity effect on *Senecio vulgaris*. No effect on population size on either species
Brevicoryne brassicae	*Brassica oleracea*								Reduced fecundity
Sitobion avenae	*Poa annua*	700 ppm	–	–	–	–	–	Díaz et al. (1998)	No effect on fecundity
Sitobion avenae	*P. trivialis*								
Sitobion avenae	*Festuca ovina*								
Sitobion avenae	*Helicotrichon pratense*								
Phyllaphis fagi	*Fagus sylvatica*	600 ppm	–	–	–	–	–	Docherty et al. (1997)	Reduced weight and number (non-significant)
Periphyllus tes-tudinaceus	*Acer pseudo-platanus*								Relative growth rates (non-significant)
Drepanosiphum platanoidis	*Acer pseudo-platanus*								Relative growth rates (non-significant)

Table 2.1 (continued)

Species	Plant-host	CO_2	O_3	T	P	D	N	References	Behavioural response
Brevicoryne brassicae	*Brassica napus*	720 ppm	–	24/20 C				Himanen et al. (2008)	The development of *Brevicoryne brassicae* (specialist) was less affected.
Myzus persicae	*Brassica napus*								The developmental time of *Myzus persicae* (generalist), was reduced and adult and progeny were smaller.
									The aphids performed equally well on Bt and non-Bt oilseed rape, which is revealing the absence of plant composition-related effects on these pests under elevated CO_2, elevated temperature or combined elevated CO_2 and temperature conditions.
Schizolachnus pineti	*Pinus sylvestris*	–	Between 1.2 × amb. and 1.7 × amb.	–	–	–	+	Holopainen et al. (1997)	As in the previous studies, aphid response to ozone was extremely variable. The expected 20–70% increase in ambient concentrations of the tropospheric ozone may in some cases enhance aphid performance, but in most cases the ozone effect on the susceptibility of conifer seedlings to sucking insect pests is not important.

Table 2.1 (continued)

Species	Plant-host	CO_2	O_3	T	P	D	N	References	Behavioural response
Cinara pinea	*Pinus sylvestris*								
Cinara pilicornis	*Picea abies*								
Cinara pilicornis	*Picea abies*	–	–	–	–	–	40, 80 and 160 ppb	Holopainen and Kössi (1998)	Elevated O_3 concentration during early shoot elongation may stimulate population development of *C. pilicornis*, but on maturing shoots, high O_3 concentration has a negative effect on aphid performance. This might be due to accelerated ageing of O_3 exposed shoots.
Rhopalosiphum padi *Sitobion avenae*		700 ppm	–	+	–	–	–	Hoover and Newman (2004)	The model suggests that, while parasitoids do have an impact on the aphid colony population dynamics, they do not fundamentally alter the aphid's response to climate change. The model predicts that for both aphids and their parasitoids, the population responses to combined effects of elevated CO_2 and temperature will be more similar to current ambient conditions than we might expect from the individual effects of CO_2 or temperature increases.

Table 2.1 (continued)

Species	Plant-host	CO_2	O_3	T	P	D	N	References	Behavioural response
Acyrthosiphon pisum	*Vicia faba*	700 ppm	–	–	–	–	–	Hughes and Bazzaz (2001)	No effect on aphid density
Aphis nerii	*Asclepias syriaca*								No effect on aphid density
Aphis oenotherae	*Oenothera biennis*								No effect on aphid density
Aulacorthum solani	*Nicotiana sylvestris*								Decreased aphid density
Myzus persicae	*Solanum dulca-mara*								Increased aphid density
Metopolophium dirhodum	*Triticum* spp. cv. "Remus"	–	–	27, 29, 31, and 33	–	–	–	Ma et al. (2004)	Increasing the temperature above 29°C decreased lifetime fecundity and longevity. Mature aphids were more sensitive to high temperatures than young aphids High temperature pulses shortened longevity and decreased lifetime fecundity
Acyrthosiphon pisum	*Vicia faba*	537 ± 77 μll^{-1}	51 ± 22 nll^{-1}		–	–	–	Mondor et al. (2005)	Different genotypes of *A. pisum* showed qualitatively different responses in population size, genotype and phenotype frequencies following exposure to a combination of elevated CO_2 and O_3

Table 2.1 (continued)

Species	Plant-host	CO_2	O_3	T	P	D	N	References	Behavioural response
Rhopalosiphum padi	*Festuca arundinacea*	700 ppm	–	–	–	–	–	Newman et al. (1999)	Smaller populations size and densities (significant)
Chaitophorus stevensis	*Populus tremuloides*	amb. and 560 ll^{-1}	amb. and 46.4–55.5 μll^{-1}	–	–	–	–	Percy et al. (2002)	Aphid abundance not affected by CO_2 due to the increase of natural enemies Aphid abundance not affected by O_3 but negatively by natural enemies. CO_2 reduced negative effects of O_3
Euceraphis betulae	*Betula pendula*	2x amb.	2x amb.	–	–	–	–	Peltonen et al. (2006)	It was found that elevated CO_2, irrespective of ozone concentration, increased the number of aphid eggs laid by one clone, but not another clone. Elevated CO_2 and O_3 affected the phenolic composition of the exudates of birch but did not correlate with the observed changes in aphid oviposition
Rhopalosiphum padi *Sitobion avenae*	*Triticum* spp.	–	–	–	+	–	–	Pons et al. (1993)	A decrease in winter rainfall could result in reduced aphid development rates

Table 2.1 (continued)

Species	Plant-host	CO_2	O_3	T	P	D	N	References	Behavioural response
Rhopalosiphum padi	*Triticum* spp.	–	–	–	–	+	–	Pons and Tatchell (1995)	Drought-stressed tillering cereals reduce the reproductive capacity of overwintering aphids
Sitobion avenae	*Triticum* spp.								
Aphis fabae	*Cardamine pratensis*	600 ppm	–	–	–	–	–	Salt et al. (1996)	No significant effects on the population sizes of either shoot- or root-feeding aphids at elevated levels of CO_2
Pemphigus populi-transversus	*Cardamine pratensis*								
Pemphigus populi-transversus	*Cardamine pratensis*	Elevated	–	–	–	–	–	Smith (1996)	No effects after initial increase in numbers
Brevicoryne brassicae	*Cardamine hirsuta*	Elevated							
Myzus persicae	*Brassica oleracea*								*B. brassicae* reared on plants grown in elevated CO_2 were larger and accumulated more fat, while there was no change in *M. persicae*. Fecundity of individual aphids increased when reared on plants grown in elevated CO_2. The differences were lost when aphids were reared in colonies.

Table 2.1 (continued)

Species	Plant-host	CO_2	O_3	T	P	D	N	References	Behavioural response
Brevicoryne brassicae *Myzus persicae*	*Brassica oleracea*	650 ppm	–	–	–	–	–	Stacey and Fellowes (2002)	CO_2 treatment influenced aphid distribution on plants. There was a lower ratio of *M. persicae*: *B. brassicae* on plants grown under elevated CO_2 conditions. The first empirical evidence that changes in host plant quality mediated by increasing levels of CO_2 can alter the outcome of interspecific competition among insect herbivores.
Macrosiphum euphorbiae	*Solanum dulcamara*	750 ppm	–	–	–	–	1–3 mmol	Sudderth et al. (2005)	Increase in aphid population size in response to elevated CO_2 or increased N availability for aphids feeding on *S. dulcamara* grown under low N conditions. No population size responses were observed for aphids infesting *A. viridis*. The maximum population growth rate that *M. euphorbiae* aphids can attain.

Table 2.1 (continued)

Species	Plant-host	CO_2	O_3	T	P	D	N	References	Behavioural response
Macrosiphum euphorbiae	*Amaranthus viridis*								
Brachycaudus helichrysi *Elatobium abietinum* *Metopolophium dirhodum* *Myzus persicae* *Sitobion avenae*	–	–	–	–	–	–		Zhou et al. (1995)	A 1 °C increase in average winter temperature advanced the migration phenology by 4–19 days depending on the species. Effects of temperature on the aphid phenology are similar between holocyclic and anholocyclic species, unlike the effects of temperature on date of the first flight record

of CO_2, an important greenhouse gas. These studies also showed that ice build-up was quite variable before the stable Holocene period (see geological time scale in Table 2.2). This was particularly true during the Quaternary Period, including the Pleistocene and Holocene Epochs. This period was characterized by repeated cycles of glaciation and periods of interglacial warming. During the Pleistocene, the main areas of glaciation were North America, Greenland, Eurasia and Antarctica. In the Holocene, climate changes were smaller in amplitude than in the Pleistocene (Easterbrook 1999). The Holocene is the interglacial period in which human civilization developed and modern plant and animal distributions stabilized (Eddy and Oeschger 1993).

The distribution patterns of animals and plants have been subjected to dramatic changes throughout time. It is known that the climate during the last ice age, was different from today (Frenzel et al. 1992), and distribution of nearly all plant and animal species was different (Huntley and Birks 1983; Hewitt 1996; Taberlet et al. 1998; Comes and Kadereit 1999; 1999). The paleontological records provide some

Table 2.2 Geological time scale (mya = million years ago); Courtesy of UC Museum of Paleontology, www.ucmp.berkeley.edu

Phanerozoic Eon (543 mya to present)	Cenozoic Era (65 mya to today)	Quaternary (1.8 mya to today) – Holocene (10,000 years to today) – Pleistocene (1.8 mya to 10,000 years) Tertiary (65–1.8 mya) – Pliocene (5.3–1.8 mya) – Miocene (23.8–5.3 mya) – Oligocene (33.7–23.8 mya) – Eocene (54.8–33.7 mya) – Paleocene (65–54.8 mya)
	Mesozoic Era (248–65 mya)	– Cretaceous (144–65 mya) – Jurassic (206–144 mya) – Triassic (248–206 mya)
	Paleozoic Era (543–248 mya)	– Permian (290–248 mya) – Carboniferous (354–290 mya)[a] – Devonian (417–354 mya) – Silurian (443–417 mya) – Ordovician (490–443 mya) – Cambrian (543–490 mya)
Precambrian Time (4,500–543 mya)	Proterozoic Era (2,500–543 mya)	Neoproterozoic (900–543 mya) Vendian (650–543 mya) Mesoproterozoic (1,600–900 mya) Paleoproterozoic (2,500–1,600 mya)
	Archaean (3,800–2,500 mya) Hadean (4,500–3,800 mya)	

[a]First fossil record of aphids on the Earth

cues about how insects have responded to past climatic changes. Most of the evidence indicates that the usual response of insects to climate change was a change in their geographical ranges (review the work of Coope), reflecting changes in the distribution of their host plants. However, these changes are constrained because tracking suitable thermal climates often involve changes in latitude and associated changes in photoperiod, which could limit species establishment. Species that change their latitude must be able to adjust to these changes. Consequently, the species that go extinct are likely to be those that are unable to adapt to the changes associated with changes in latitudinal distribution, and species that have limited mobility, like apterous forms. In aphids, there is no evidence of wind or water-assisted dispersal of apterous forms, although eggs attached to vegetation may be dispersed over short distances (Fahnestock et al. 2000).

Insects are particularly vulnerable to changes in temperature, availability of water, and air and water chemistry because of their relatively large surface area to volume ratios. No individual species is capable of tolerating the entire range of tropical to arctic temperatures or desert to moisture conditions (Schowalter 2000). Due to their small size, short life spans and high reproductive rates, insects can minimize time lags between environmental change and population adjustment to new conditions.

Despite the fact that the fate of insects is a better indicator of environmental change than that of larger or longer-lived organisms (Schowalter 2000) there are very few such studies on invertebrates (Cooper et al. 1995).

The fact that most fossils were not given names of extant species was because workers believed that most organisms worldwide became extinct during the Pleistocene and the species living today evolved after this extinction (Elias 1994). Some fossil beetles were classified as extant species, and it was not until the mid-1900s that Carl Lindroth and Russell Coope argued that most Quaternary beetles still survive today (Elias 1994). Lindroth revised past studies of fossil insects in Sweden during the 1940s, giving them extant names and pointing out the taxonomic errors of others. In 1955 Coope collected fossil beetles at Upton Warren near Birmingham, UK. He made comparisons with museum specimens and inferred that all the specimens could be identified as modern species. Coope published over one hundred papers on fossil beetles and their paleoenvironmental significance. These characteristics of insects make them good indicators of how organisms have responded to climatic changes in the past.

2.3 Aphids as Fossils

Aphids have very remarkable multiple-generation complex life cycles, similar to those of some animal parasites and rust fungi (Moran 1994). The life cycle complexity of aphids is associated with cyclical parthenogenesis and host plant alternation. Some authors argue that complex life cycles are important determinants of success when colonizing new ecological niches.

Fossil evidence indicates that Aphidoidae appeared 280 million years ago (see Table 2.2), in the Carboniferous Era (Heie 1967; Dixon 1998). Aphid fossils occur in Canadian amber (75–80 million years old), Baltic amber (Eocene, 35–45 million years old) and clay and other sediments from the Triassic to the Pleistocene. There is some evidence that aphids are capable of adapting to climate change because fossil ones are often still extant, which indicates that extreme climatic events in the past did not always result in major evolutionary changes or extinction; Fig. 2.1 summarizes the occurrence in time of aphid fossils and their taxonomic status based on Blackman and Eastop (1984, 1994).

Aphids appeared in the fossil record before the period of greatest climate instability, and even before the decline in gymnosperm and appearance of angiosperm plants in the Cretaceous Period. Heie (1987, 1996) suggested that the presence of several specialized groups in upper Cretaceous Canadian amber, but not later in the fossil record, is evidence of an extinction event near the Cretaceous-Tertiary boundary. This large-scale extinction was hypothesized to be connected with the extinction of gymnosperms, perhaps driven by the diversification of angiosperms in the Middle to Late Cretaceous (Fig. 2.2; Heie 1987; Shaposhnikov 1987; Heie and Pike 1992, 1996).

Therefore, aphids have already faced declines in the abundance and extinction of their host plants and periods of abrupt climate change in the past. But the question addressed here is how will aphids deal with the current predicted changes in climate? What are the major differences between today's world and the pre-historic world? The current situation is thought to be mainly due to man. The development of agriculture resulted in the homogenization of landscapes, which contributed to the establishment of agriculture pests. More recently, the industrial revolution has changed not only the environment, but also our way of living, with the development of transports and roads to speed our lives. Cities have grown and more natural landscapes were cut-off to allow mankind to grow. These roads and cities have contributed to fragment natural landscapes and even in some touristic destinations, agricultural fields have been replaced by houses and roads. During the Ice Age many species survived by migrating to appropriate habitats. Nowadays, such migrations would be much more difficult or even impossible, because they would have to cross roads, agricultural areas, and all kinds of human settlements.

Habitat loss is the most significant cause of species extinctions (Wilcove et al. 1998). Fragmentation of the remaining habitat often occurs during habitat loss, and this has been linked to decreased species diversity in a variety of taxonomic groups (Crooks 2002; Drinnan 2005). The synergistic, or combined, effects of habitat fragmentation and climate change can represent one of the most potentially serious global change problems. In the case of aphids this will be problematic for species that are not agricultural pests, because natural landscapes will be more scattered and more difficult to find. For crop pests these problems would be much smaller. Also, some of the empirical works that had shown population increases or increase fitness were performed using aphids that are pest crops.

			Aphididae																					Ad	Py	El	Me	Pa	Ta	Sh	Si	Ge	Cr	Un
Period	Epoch/ Division	Began MYA	Er		An	Ho	Mi	Dr							Ct	Th	Gr	Ap	La		Un	Ca	Ov											
			Er	Pr				Ne	Ph	Dr	Li	Sa	Ch	Un	†				Eu	La		†	†			†	†	†	†	†	†	†	†	†
Quaternary	Pleistocene	2											x											x										
Tertiary	Pliocene	5	x										x								x			x		x								
	Miocene	24	x								x						x		x	x					x									
	Oligocene	36			x	x	x			x				x			?				x					x								x
	Eocene *	57	x	x	x			?	x	x	x	x		x		x		x																
	Paleocene	65																																
Cretaceous	Upper **	~80					?	x			?				x			x				x				x	x	x	x	x				x
	Middle													?																				
	Lower	146				?								?				?				x	x					x		x	x	x		x
Jurassic	Upper													?							x		x											
	Middle																																	
	Lower	208																																
Triassic	Upper																																x	x
	Middle																																x	x
	Lower	245																																

* Baltic amber **Canadian & Tajmyr amber ☐ Extant families † Extinct taxon

Fig. 2.1 Summary of known aphid fossil, based on Heie and Wegierek (1998) and Heie and Penalver (1999). The geological time scale is based on Harland et al. (1990). Ad=Adelgidae, Ap=Aphididae, An=Anoeciinae, Ca=Canadaphidinae, Ch=Chaitophorini, Cr=Creaphididae, Ct=Cretamyzinae, Dr=Drepanosiphinae/ini, Er=Eriosomatinae/ini, El=Elektraphididae, Eu=Eulachnini, Ge=Genephididae, Gr=Greenideinae, Ho=Hormaphidinae, La=Lachninae/ini, Li=Lizerini, Me=Mesozoicaphididae, Mi=Mindarinae, Ne=Neophyllaphidini, Ov=Oviparosiphinae, Pa=Palaeosphididae, Ph=Phyllaphidina, Pr=Prociphilina, Py=Phylloxeridae, Sa=Saltusaphidini, Sh=Shaposhnikovidae, Si=Sinaphididae, Ta=Tajmyraphididae, Th=Thelaxinae, Un=Unknown, ?= placement of fossil uncertain (adapted from Von Dohlen and Moran 2000, with permission of the authors)

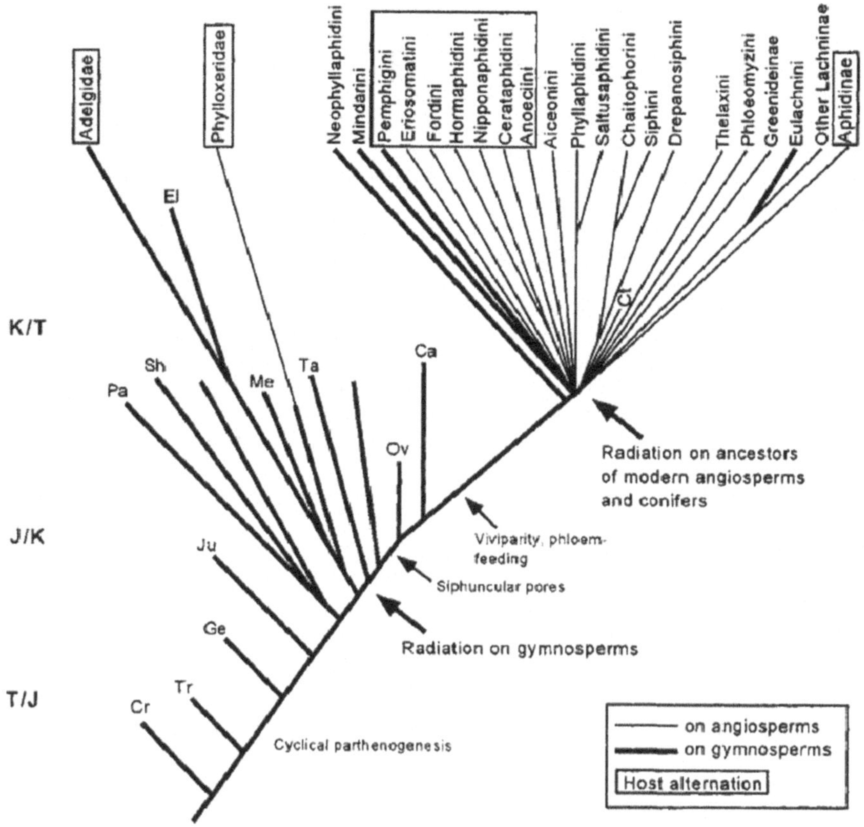

Fig. 2.2 Hypothetical evolutionary history of aphids. For abbreviations of extinct taxa see Fig. 2.1. The category "on gymnosperms" includes tribes considered to be primitively associated with gymnosperms. K/T=Cretaceous/Tertiary boundary, J/K= Jurassic/Cretaceous boundary, T/J= Triassic/Jurassic boundary (adapted from Von Dohlen and Moran 2000, with permission of the authors)

Acknowledgements I would like to thank to António Pombinho for his support and critical reading of the early versions of this manuscript. The paper also greatly benefited from discussions with Professor Tony Dixon. The work was supported by the grant No. LC 06073 of the MSMT.

References

Andrewartha HG, Birch LC (1954) The distribution and abundance of animals. University of Chicago Press, Chicago, IL

Awmack CM, Harrington R (2000) Elevated CO_2 affects the interactions between aphid pests and host plant flowering. Agric For Entomol 2:57–61

Awmack CS, Harrington R, Leather SR (1997) Host plant effects on the performance of the aphid *Aulacorthum solani* (Kalt.) (Homoptera: Aphididae) at ambient and elevated CO_2. Glob Chang Biol 3:545–549

Awmack CS, Harrington R, Leather SR, Lawton JH (1996) The impacts of elevated CO_2 on aphid-plant interactions. Asp Appl Biol 45:317–322

Awmack CS, Harrington R, Lindroth RL (2004) Aphid individual performance may not predict population responses to elevated CO_2 or O_3. Glob Chang Biol 10:1414–1423

Bezemer TM, Knight KJ, Newington JE, Jones TH (1999) How general are aphid responses to elevated atmospheric CO_2? Ann Entomol Soc Am 92:724–730

Blackman RL, Eastop VF (1984) Aphids on the world's crops: an identification guide. John Wiley and Sons, Chichester, UK

Blackman RL, Eastop VF (1994) Aphids on the world's trees: an identification and information guide. CAB International, Wallingford

Comes HP, Kadereit JW (1999) The effect of Quaternary climatic changes on plant distribution and evolution. Trends Plant Sci 3:432–438

Coope GR (1995) Insect faunas in ice age environments: why so little extinction? In: Lawton JH, May RM (eds) Extinction rates. Oxford University Press, Oxford, pp 55–74

Cooper SJ, Ibrahim KM, Hewitt GM (1995) Postglacial expansion and genome subdivision in the European grasshopper *Chorthippus parallelus*. Mol Ecol 4:49–60

Crawley MJ (1983) Herbivory, the dynamics of animal-plant interaction. University of California Press, Berkeley

Crooks KR (2002) Relative sensitivities of mammalian carnivores to habitat fragmentation. Conserv Biol 16:488–502

Davis MB, Zabinski C (1992) Changes in geographical range resulting from greenhouse warming: effects on biodiversity in forests. In: Peters RL, Lovejoy TE (eds) Global warming and biological diversity. Yale University Press, New Haven, Connecticut, pp 297–308

Dixon AFG (1985) Aphid ecology. Chapman & Hall, Glasgow

Dixon AFG (1998) Aphid ecology: an optimization approach. Chapman and Hall, London

Docherty M, Wade FA, Hurst DK, Whittaker JB, Lea PJ (1997) Responses of tree sap-feeding herbivores to elevated CO_2. Glob Chang Biol 3:51–59

Drinnan IN (2005) The search for fragmentation thresholds in a Southern Sydney suburb. Ecol Model 124:339–349

Díaz S, Fraser LH, Grime JP, Falczuk V (1998) The impact of elevated CO_2 on plant-herbivore interactions: experimental evidence of moderating effects at the community level. Oecologia 117:177–186

Easterbrook DJ (1999) Surface processes and landforms. Prentice Hall, New Jersey

Eddy JA, Oeschger H (1993) Global changes in the perspective of the past. John Wiley & Sons, New York

Elias SA (1994) Quaternary insects and their environments. Smithsonian Institution Press, Washington

Fahnestock JT, Povirk KL, Welker JM (2000) Ecological significance of litter redistribution by wind and snow in arctic landscapes. Ecography 23:623–631

Flynn DFB, Sudderth EA, Bazzaz FA (2006) Effects of aphid herbivory on biomass and leaf-level physiology of *Solanum dulcamara* under elevated temperature and CO_2. Environ Exp Bot 56:10–18

Frenzel B, Pécsi M, Velichko AA (1992) Atlas of paleoclimates and paleoenvironments of the Northern Hemisphere. Late Pleistocene-Holocene. Gustav Fischer, Stuttgart

Hadly EA (1997) Evolutionary and ecological response of pocket gopher (*Thomomys talpoides*) to Late-Holocene climatic change. Biol J Linn Soc 60:277–296

Harland WB, Armstrong RL, Cox AV, Craig LE, Smith AG, Smith DG (1990) A geologic time scale 1989. Cambridge University Press, Cambridge, UK

Heie OE (1967) Studies on fossil aphids (Homoptera: Aphidoidea), especially in the Copenhagen collection of fossils in Baltic amber. Spolia Zoologica Musei Hauniensis 26:1–274

Heie OE (1987) Morphological structures and adaptation. In: Minks AK, Harrewijn P (eds) Aphids: their biology, natural enemies and control. Elsevier, Amsterdam, pp 393–400

Heie OE (1996) The evolutionary history of aphids and a hypothesis on the coevolution of aphids and plants. Bollettino di Zoologia Agraria e di Bachicoltura 28:149–155

Heie OE, Peñalver E (1999) *Palaeophylloxera* nov. gen., the first fossil specimen of the family Phylloxeridae (Hemiptera: Phylloxeroidea); Lower Miocene of Spain. Geobios 32:593–597

Heie OE, Pike EM (1992) New aphids in Cretaceous amber from Alberta (Insecta, Homoptera). Can Entomol 124:1027–1053

Heie OE, Pike EM (1996) Reassessment of the taxonomic position of the fossil aphid family Canadaphididae based on two additional specimens of *Canadaphis carpenteri* (Hemiptera, Aphidinea). Eur J Entomol 93:617–622

Heie OE, Wegierek P (1998) A list of fossil aphids (Homoptera, Aphidinea). Ann Upper Silesian Museum (Entomol) 8–9:159–192

Hewitt GM (1996) Some genetic consequences of ice ages, and their role in divergence and speciation. Biol J Linn Soc 58:247–276

Himanen SJ, Nissinen A, Dong W-X, Nerg A-M, Stewart CN Jr, Poppy GM, Holopainen JK (2008) Interactions of elevated carbon dioxide and temperature with aphid feeding on transgenic oilseed rape: are *Bacillus thuringiensis* (Bt) plants more susceptible to nontarget herbivores in future climate? Glob Chang Biol 14:1–18

Holopainen JK (2002) Aphid response to elevated ozone and CO_2. Entomol Exp Appl 104: 137–142

Holopainen JK, Kainulainen P, Oksanen J (1997) Growth and reproduction of aphids and levels of free amino acids in Scots pine and Norway spruce in an open-air fumigation with ozone. Glob Chang Biol 3:139–147

Holopainen JK, Kössi S (1998) Variable growth and reproduction response of the spruce shoot aphid, *Cinara pilicornis*, to increasing ozone concentrations. Entomol Exp Appl 87:109–113

Hoover JK, Newman JA (2004) Tritrophic interactions in the context of climate change: a model of grasses, cereal aphids and their parasitoids. Glob Chang Biol 10:1197–1208

Houghton JT, Ding Y, Griggs DJ, Noguer M, van der Linden PJ, Xiaosu D, Maskell K, Johnson CA (2001) Climate change 2001: the scientific basis. Cambridge University Press, Cambridge

Hughes L, Bazzaz FA (2001) Effects of elevated CO_2 on five plant-aphid interactions. Entomol Exp Appl 99:87–96

Huntley B, Birks HJB (1983) An atlas of past and present pollen maps of Europe: 0–13,000 years ago. Cambridge University Press, Cambridge

IPCC (2001) Summary for policymakers. Report of working group I of the intergovernmental panel on climate change. Intergovernmental Panel on Climate Change. http://www.ipcc.ch/pub/spm22-01.pdf

Körner C (2000) Biosphere responses to CO_2 enrichment. Ecol Appl 10:1590–1619

Körner C (2003) Ecological impacts of atmospheric CO_2 enrichment on terrestrial ecosystems. R Soc Lond Trans A 361:2023–2041

Ma C-S, Hau B, Poehling H-M (2004) Effects of pattern and timing of high temperature exposure on reproduction of the rose grain aphid, *Metopolophium dirhodum*. Entomol Exp Appl 110: 65–71

Martin TE (1998) Are microhabitat preferences of coexisting species under selection and adaptive? Ecology 79:656–670

Mondor EB, Tremblay MN, Awmack CS, Lindroth RL (2005) Altered genotypic and phenotypic frequencies of aphid populations under enriched CO_2 and O_3 atmospheres. Glob Chang Biol 11:1990–1996

Moran NA (1994) Adaptation and constraint in the complex life cycles of animals. Ann Rev Ecol Evol Syst 25:573–600

Newman JA, Gibson DJ, Hickam E, Lorenz M, Adams E, Bybee L, Thompson R (1999) Elevated carbon dioxide results in smaller populations of the bird cherry-oat aphid *Rhopalosiphum padi*. Ecol Entomol 24:486–489

Parmesan C (1996) Climate and species' range. Nature 382:765–766

Peltonen PA, Julkunen-Tiitto R, Vapaavuori E, Holopainen JK (2006) Effects of elevated carbon dioxide and ozone on aphid oviposition preference and birch bud exudate phenolics. Glob Chang Biol 12:1670–1679

Percy KE, Awmack CS, Lindroth RL, Kubiske ME, Kopper BJ, Sebrands JGI, Pregitzer KS, Hendrey GR, Dickson RE, Zak DR, Oksanen E, Sober J, Harrington R, Karnosky DF (2002) Altered performance of forest pests under atmospheres enriched by CO_2 and O_3. Nature 420:403–407

Pickett STA, White PS (1985) The ecology of natural disturbance and patch dynamics. Academic Press, San Diego

Pons X, Comas J, Albajes R (1993) Overwintering of cereal aphids (Homoptera: aphididae) on durum wheat in a Mediterranean climate. Environ Entomol 22:381–387

Pons X, Tatchell GM (1995) Drought stress and cereal aphid performance. Ann Appl Biol 126: 19–31

Prather M, Ehhalt D (2001) Atmospheric chemistry and greenhouse gases. In: Houghton JT, Ding Y, Griggs DJ, Noguer M, Van der Linden PJ, Dai X, Maskell K, Johnson CA (eds) Climate change 2001: the scientific basis. Cambridge University Press, Cambridge, pp 239–287

Prentice IC (2001) The carbon cycle and atmospheric carbon dioxide. In: Houghton JT, Ding Y, Griggs DJ, Noguer M, Van der Linden PJ, Dai X, Maskell K, Johnson CA (eds) Climate change 2001: the scientific basis. Cambridge University Press, Cambridge, UK, pp 183–237

Root TL (1988) Environmental factors associated with avian distributional limits. J Biogeogr 15:489–505

Rubenstein DI (1992) Global warming and biological diversity. Yale University Press, New Haven

Salt DT, Fenwick P, Whittaker JB (1996) Interspecific herbivore interactions in a high-CO_2 environment: root and shoot aphids feeding on Cardamine. Oikos 77:326–330

Schowalter TD (2000) Insect ecology: an ecosystem approach. Academic Press, San Diego, California

Shaposhnikov GC (1987) Evolution of aphids in relation to evolution of plants. In: Minks AK, Harrewijn P (eds) Aphids: their biology, natural enemies and control. Elsevier, Amsterdam, pp 409–414

Singer MC, Thomas CD (1996) Evolutionary responses of a butterfly metapopulation to human- and climate-caused environmental variation. Am Nat 148:S9–S39

Smith H (1996) The effects of elevated CO_2 on aphids. Antenna 20:109–111

Stacey DA, Fellowes MDE (2002) Influence of elevated CO_2 on interspecific interactions at higher trophic levels. Glob Chang Biol 8:668–678

Sudderth EA, Stinson KA, Bazzaz FA (2005) Host-specific aphid population responses to elevated CO_2 and increased N availability. Glob Chang Biol 11:1997–2008

Taberlet P, Fumagalli L, Wust-Saucy A-G, Cosson J-F (1998) Comparative phylogeography and postglacial colonization routes in Europe. Mol Ecol 7:453–464

Theurillat P, Guisan A (2001) Potential impact of climate change on vegetation in the European Alps: a review. Climatic Change 50:77–109

Von Dohlen CD, Moran NA (2000) Molecular data support a rapid radiation of aphids in the Cretaceous and multiple origins of host alternation. Biol J Linn Soc 71:689–717

Whittaker JB (1999) Impacts and responses at population level of herbivorous insects to elevated CO_2. Eur J Entomol 96:149–156

Wilcove DS, Rothstein D, Dubow J, Phillips A, Losos E (1998) Quantifying threats to imperiled species in the United States. BioScience 48:607–615

Williams RS, Lincoln DE, Norby RJ (2003) Development of gypsy moth larvae feeding on red maple saplings at elevated CO_2 and temperature. Oecologia 137:114–122

Woodward FI (1987) Climate and plant distribution. Cambridge University Press, Cambridge

Zhou X, Harrington R, Woiwod IP, Perry JN, Bale JS, Clark SJ (1995) Effects of temperature on aphid phenology. Glob Chang Biol 1:303–313

Chapter 3
Trends in the Timings of the Start and End of Annual Flight Periods

Richard Harrington and Suzanne Clark

Abstract Younger plants tend to be more susceptible than older plants to damage caused both by feeding and by viruses. The impact of any change in aphid phenology will depend on the direction and extent of changes in host plant phenology. If changes in host plant phenology exactly mirror changes in aphid phenology, damage may be unchanged. Higher trophic levels are also important considerations. Changes in the synchrony of aphids and their natural enemies, or their competitors, will be determinants of aphid dynamics and damage. This chapter deals only with the aphid component, and utilises the long-term data from the UK network of suction traps operated by the Rothamsted Insect Survey.

Keywords Aphids · Synchrony · Suction traps · Rothamsted Insect Survey

3.1 Introduction

Aphid phenology matters. Younger plants tend to be more susceptible than older plants to damage caused both by feeding and by viruses. For example, Smith and Hallsworth (1990) showed that loss of sugar yield in sugar beet plants was greater when aphid invasion occurred earlier. In the case of potatoes, viruses inoculated later in crop development are less likely to infect tubers because of the gradual breakdown of the phloem transport system in maturing plants (Sigvald 1985). Both of these crops are hosts of *Myzus persicae*, one of the most damaging aphids of UK and European agriculture. Thus, the impact of any change in aphid phenology will depend on the direction and extent of changes in host plant phenology. If changes in host plant phenology exactly mirror changes in aphid phenology,

R. Harrington (✉)
Rothamsted Research, Hertfordshire AL5 2JQ, UK
e-mail: richard.harrington@bbsrc.ac.uk

P. Kindlmann et al. (eds.), *Aphid Biodiversity under Environmental Change*,
DOI 10.1007/978-90-481-8601-3_3,

damage may be unchanged. Higher trophic levels are also important considerations. Changes in the synchrony of aphids and their natural enemies, or their competitors, will be determinants of aphid dynamics and damage. This chapter deals only with the aphid component, and utilises the long term data from the UK network of suction traps operated by the Rothamsted Insect Survey (Harrington and Woiwod 2007).

Such statistical approaches have their advantages and limitations. A severe limitation is usually data availability, but that is not the case here. Relationships found between aphid phenology and temperature hold in spite of the influences of other abiotic (e.g., climatic, pollutant) and biotic (e.g., host plant, natural enemy) factors; however, they do not quantify the effects of these other factors, which are just part of the noise (error) in the relationships found. Prediction, especially longer term in relation to climate change, carries a risk, in that the position in multidimensional space of the values of relevant variables in the future may be outside the range of those used to construct the relationships. Combinations of factors not previously experienced may lead to a weakening of relationships in the data. However, an alternative method of prediction involving the construction of process-based models would be hugely complex and the models difficult to parameterise.

Thus, this chapter makes use of the best available long-term data in the world on aphid dynamics, whilst acknowledging some of the risks involved in interpretation and prediction. At the very least, relationships found can lead to hypotheses as to their cause, which can then be tested experimentally. This approach has led to useful insights on numerous occasions. The objective here is to examine the timing of the start of the annual aphid flight season in relation to winter temperature, and to see if trends with time (year) are discernible in the timing of the start and the end of the aphid flight season. We investigate whether there is any evidence for (i) the start of the flight season becoming earlier as winters get warmer; (ii) this relationship being stronger for aphids that tend to be continuously parthenogenetic (anholocyclic or androcyclic) in the area of interest than for those which pass the winter as an egg (holocyclic), because mobile stages are far less cold hardy than eggs (Leather et al. 1993) but, in the case of most aphids, do not enter diapause; and (iii) a trend towards migration beginning earlier and (iv) ending later, in more recent years, because average temperatures have risen.

3.2 Methods

The aphid data come from the 12.2 m suction trap (Macaulay et al. 1988) network operated by the Rothamsted Insect Survey (Harrington and Woiwod 2007). Although the trap network now covers much of Europe (Harrington et al. 2004), this study is confined to 13 traps in the UK, which are still operated today and have at least 25 years of data (Table 3.1, Fig. 3.1).

The dates of first and last record of 20 common species (Table 3.2) in the trap at Rothamsted were recorded for each year. The dates of first and last record of one

Table 3.1 Location and dates of operation of the 13 suction traps used in this study

Site	Latitude	Longitude	Weather station	Years analysed
Ayr	55.477°N	4.567°W	Adjacent	1975–2005[a]
Broom's Barn	52.260°N	0.570°E	Adjacent	1968–2005
Dundee	56.457°N	3.069°W	Adjacent	1968–2005
East Craigs	55.949°N	3.312°W	Adjacent	1969–2005[b]
Hereford	52.125°N	2.637°W	Adjacent	1972–2005[c]
Kirton	52.937°N	0.070°W	Adjacent	1980–2005
Newcastle	55.213°N	1.682°W	Adjacent	1968–2005
Preston	53.854°N	2.763°W	Hazelrigg (17 km)	1975–2005
Rothamsted	51.807°N	0.356°W	Adjacent	1968–2005[d]
Silwood	51.408°N	0.641°W	Hurley (19 km)	1968–1989
			Adjacent	2000–2005
Starcross	50.628°N	3.454°W	Yeovilton (69 km)	1970–2005[e]
Writtle	51.733°N	0.427°E	Adjacent	1975–2005
Wye	51.185°N	0.939°E	Adjacent	1968–2005

[a] No temperature data for 1991 and 1993; no flight data for 1991
[b] No flight data for 1987
[c] No temperature data for 1972
[d] No flight data for *A. solani* in 1972, 1977, 1979, 1981, 1986, 1997
[e] No temperature data for 2003–2005

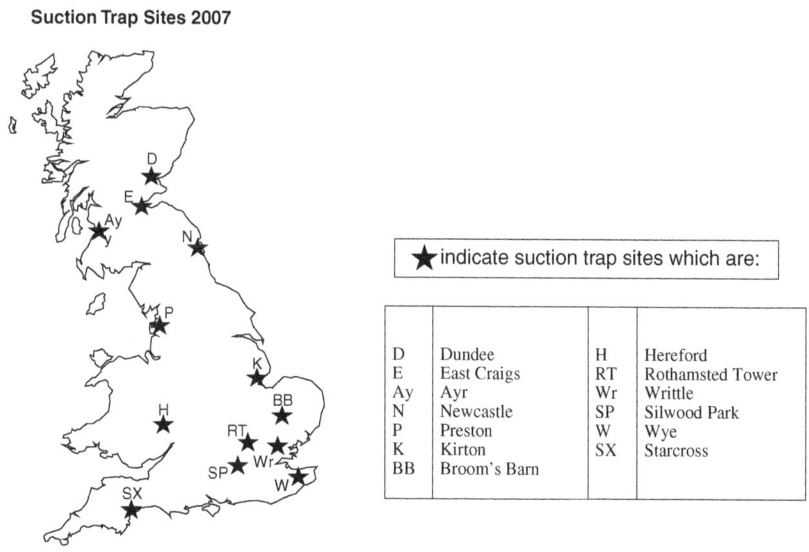

Fig. 3.1 The 13 UK suction trap sites from which data were analysed

Table 3.2 For 20 aphid species at Rothamsted, the % variance accounted for (%var) by, and significance of (P), linear regressions relating Julian date of first suction trap record to January–February mean screen temperature and to year, and Julian date of last suction trap record to year. All analyses before removal of any outliers (see text)

	1st record vs Jan–Feb temp		1st record vs year		Last record vs year	
	%var	P	%var	P	%var	P
Mainly anholocyclic/androcyclic						
Brevicoryne brassicae	67	<0.001	14	<0.05	18	<0.01
Elatobium abietinum	25	<0.001	36	<0.001	6	>0.05
Macrosiphum euphorbiae	30	<0.001	21	<0.01	85*	>0.05*
Myzus ascalonicus	28	<0.001	0	>0.05	96*	<0.05*
Myzus persicae	80	<0.001	17	<0.01	7	>0.05[a]
Sitobion avenae	64	<0.001	17	<0.01	0	>0.05
Mainly holocyclic						
Aphis fabae	24	<0.001	12	<0.05	29	<0.001
Aulacorthum solani	13	<0.05	1	>0.05	94*	>0.05*
Drepanosiphum platanoidis	18	<0.01	42	<0.001	1	>0.05
Hyalopterus pruni	9	<0.05	8	<0.05	23	<0.01
Hyperomyzus lactucae	35	<0.001	11	<0.05	74*	>0.05*
Metopolophium dirhodum	26	<0.001	0	>0.05[b]	88*	>0.05*
Nasonovia ribisnigri	0	>0.05	0	>0.05	83*	<0.05*
Phorodon humuli	40	<0.001	27	<0.001	3	>0.05
Rhopalosiphum insertum	16	<0.01	45	<0.001	7	<0.05
Sitobion fragariae	18	<0.01	15	<0.01	66*	>0.05*
Both options						
Acyrthosiphon pisum	21	<0.01	10	<0.05	2	>0.05
Brachycaudus helichrysi	48	<0.001	30	<0.001	14	<0.05
Cavariella aegopodii	35	<0.001	29	<0.001	84*	>0.05*
Rhopalosiphum padi	52	<0.001	9	<0.05	0	>0.05

*Based on two parallel lines model, P gives significance of slope parameter (t-test)
[a]P= 0.06 before and 0.37 after removal of two outliers (%var = 0)
[b]P<0.05 after removal of one outlier (%var = 9)

species, *Myzus persicae*, were recorded at all 13 traps for each year. These dates are determined partly by the innate phenological cycle, and partly by the abundance, of each species in each year (Fig. 3.2).

Temperature data come from weather stations sited near to the suction traps (Table 3.1). Based on previous studies (Harrington et al. 1990; Fleming and Tatchell 1995), but with the benefit of many more years' data, mean screen temperature for January and February was chosen as an explanatory variable for date of first record.

Simple linear regression was used to relate:

(i) date of first record to winter temperature;
(ii) date of first record to year; and
(iii) date of last record to year.

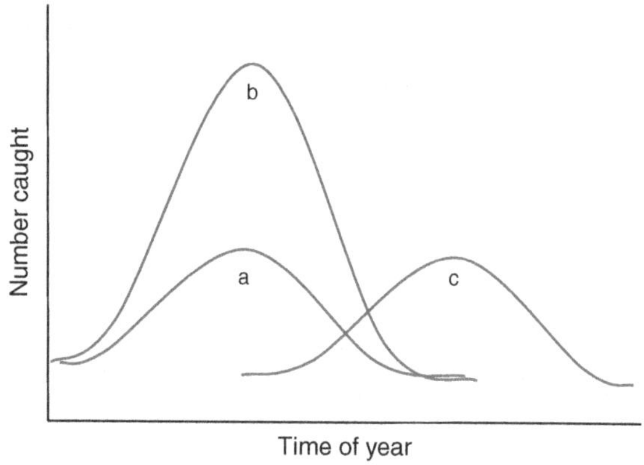

Fig. 3.2 Hypothetical phenology and abundance curves. Earlier first records will occur for *a* compared to *c* and for *b* compared to *a*

At three sites, temperature data were not available in certain years (Ayr 1991 and 1993; Hereford 1972; Starcross 2003–2005), at Ayr in 1991 and East Craigs in 1987 no *M. persicae* were caught and at Rothamsted in 6 years no *A. solani* were caught. These site-years are hence omitted from some analyses. In cases where there were clear outliers (sites with unusually early or late first or last records), analyses were repeated with the outliers excluded to check their effect on the results. In some of the analyses of date of last record, there was clearly a set of years when the last record was comparatively early and a set when it was late. In these cases a parallel model analysis (linear regression with groups) was done to compare fitted lines for each set.

3.3 Results

3.3.1 Date of First Record in Relation to Winter Temperature

Of the 20 species examined from the Rothamsted trap, higher winter temperature led to significantly ($P<0.05$) earlier first records for 19 and to significantly later first record for none (Table 3.2, Fig. 3.3). The mean advance in the date of first record of the 19 species was 8 days per °C rise in January–February temperature (range 2 days in *Hyalopterus pruni* to 19 days in *Brevicoryne brassicae*). Removal of the apparent outliers for *Aulacorthum solani*, *Metopolophium dirhodum*, *Myzus ascalonicus* and *Sitobion fragariae* (Fig. 3.3) decreased the negative slope in each case (by a maximum of *c.* 2 days per °C for *A. solani*) but did not change the overall conclusions regarding the presence of a relationship between first record and temperature. Removal of the apparent outlier for *Nasonovia ribisnigri* increased the negative slope by *c.* one day per °C but again did not change the overall conclusion of no relationship between first record and temperature.

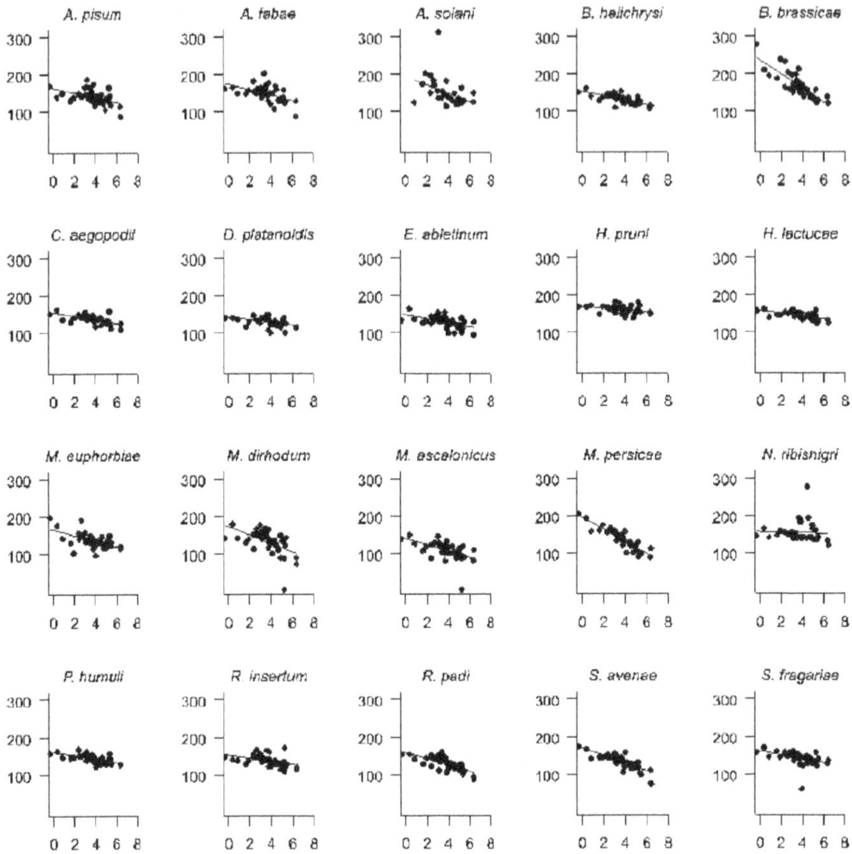

Fig. 3.3 Annual Julian date of first suction trap record (y-axis) against mean screen temperature for January and February (°C, x-axis) for 20 aphid species at Rothamsted. Regression lines shown are from analyses before removing outliers for *A. solani*, *M. dirhodum*, *M. ascalonicus*, *S. fragariae* and *N. ribisnigri* (see text). Only the line for *N. ribisnigri* has a non-significant slope ($P>0.05$)

The fits of the regressions tended to be better for those species that overwinter mainly in the mobile stages in the Rothamsted area than for those that usually overwinter as eggs. For the six species that are largely anholocyclic or androcyclic at Rothamsted, the pooled variance accounted for is 55%. For the ten species that are largely holocyclic at Rothamsted, the pooled variance accounted for is 17%. For the four species with both life cycle options open, the figure is 36%.

For *Myzus persicae*, date of first record advanced significantly ($P<0.05$) with higher winter temperature at 12 of 13 sites and became significantly later at none (Table 3.3, Fig. 3.4). The mean advance in the date of first record at the 12 sites was 12 days per °C rise in January–February temperature (range 8 days at Silwood to 16 days at Newcastle).

Table 3.3 For *Myzus persicae* at 13 sites, the % variance accounted for (%var) by, and significance of (P), linear regressions relating Julian date of first suction trap record to January–February mean screen temperature and to year, and Julian date of last suction trap record to year. All analyses before removal of any outliers (see text)

	1st record vs Jan–Feb temp		1st record vs year		Last record vs year	
	%var	P	%var	P	%var	P
Ayr	39	<0.001	0	>0.05	1	>0.05
Broom's Barn	69	<0.001	16	<0.01	0	>0.05
Dundee	32	<0.001	12	<0.05	0	>0.05
East Craigs	54	<0.001	1	>0.05	0	>0.05
Hereford	60	<0.001	9	>0.05	0	>0.05
Kirton	64	<0.001	15	<0.05	4	>0.05
Newcastle	68	<0.001	18	<0.01	0	>0.05
Preston	1	>0.05	0	>0.05	1	>0.05
Rothamsted	80	<0.001	17	<0.01	7	>0.05
Silwood	28	<0.01	7	>0.05	12	<0.05
Starcross	51	<0.001	2	>0.05	3	>0.05
Writtle	47	<0.001	31	<0.001	0	>0.05
Wye	42	<0.001	11	<0.05	0	>0.05

3.3.2 Date of First Record in Relation to Year

Of the 20 species examined from the Rothamsted trap, the first record of 16 became significantly (P<0.05) earlier with year and none significantly later (Table 3.2, Fig. 3.5). The mean advance in the 16 species was 0.72 days per year (range 0.30 days in *H. pruni* to 1.25 days in *B. brassicae*). Slopes appeared not to depend on life cycle type (Fig. 3.5). Removal of the outliers for *A. solani*, *M. ascalonicus*, *S. fragariae* and *N. ribisnigri* (as for winter temperature above) affected the slopes by less than 0.3 days per year and generally made no difference to the overall conclusions. However, after removal of the outlier for *M. dirhodum* a significant relationship (P<0.05) was achieved with a slope of –0.78 days per year.

For *M. persicae*, date of first record advanced significantly with year at seven of 13 sites and became significantly later at none (Table 3.3, Fig. 3.6). The mean advance in the date of first record at the seven sites was 1.13 days per year (range 0.90 days at Wye to 1.57 days at Writtle).

3.3.3 Date of Last Record in Relation to Year

Of the 20 species examined from the Rothamsted trap, the last record of seven became significantly (P<0.05) earlier with year and none significantly later (Table 3.2, Fig. 3.7). The mean advance in the seven species was 0.79 days per year (range 0.43 days in *Brachycaudus helichrysi* to 1.37 days in *Aphis fabae*). Slopes appeared not to depend on life cycle type (Fig. 3.7). Removal of the outlier

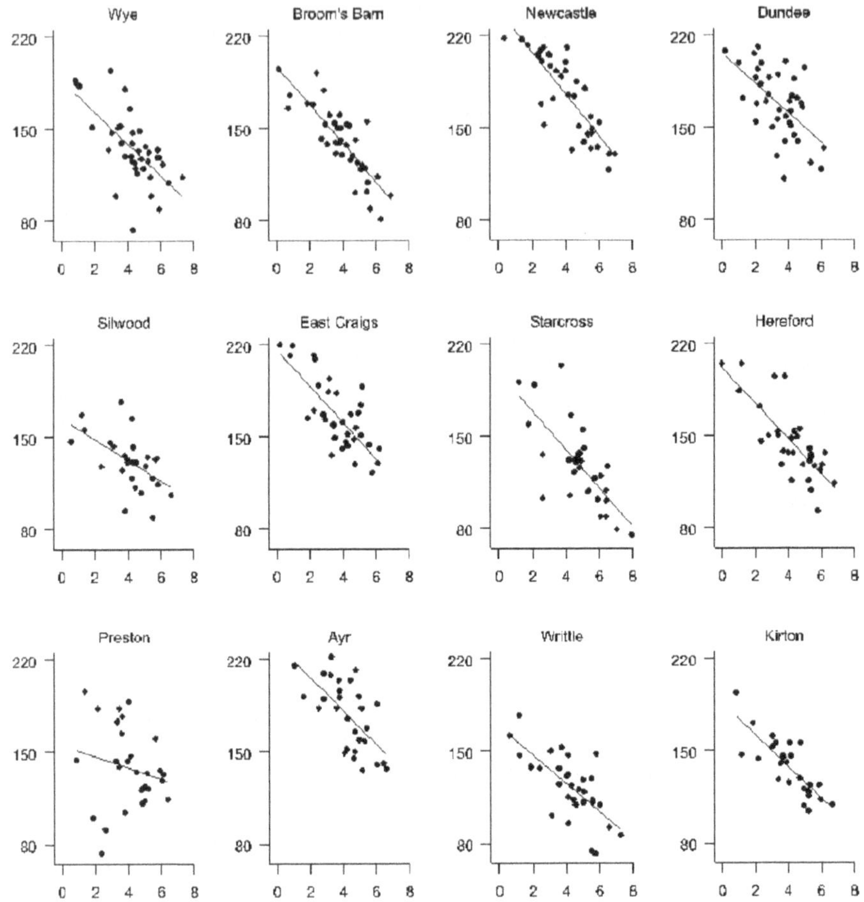

Fig. 3.4 Annual Julian date of first suction trap record (y-axis) against mean screen temperature for January and February (°C, x-axis) for *Myzus persicae* at 12 suction trap sites with regression lines superimposed (Rothamsted shown in Fig. 3.3). Only the line for Preston has a non-significant slope (P>0.05)

for *Drepanosiphum platanoidis* decreased the slope slightly but did not affect the overall conclusion of no relationship. However, removal of the two outliers for *M. persicae* resulted in a much shallower, strongly non-significant slope (–0.23 days per year) compared to the first analysis which showed weak evidence of a relationship (P=0.056). Distinct groups of years with early (before Julian date 240) and late last records were seen for *A. solani*, *Cavariella aegopodii*, *Hyperomyzus lactucae* (Julian date 247), *Macrosiphum euphorbiae*, *M. dirhodum*, *M. ascalonicus*, *N. ribisnigri* and *S. fragariae* (Fig. 3.7). A parallel lines model (i.e., different intercepts, same slope) always gave the most sufficient description of the early and late last records. The slopes were significant (P<0.05) only for *M. ascalonicus* and *N. ribisnigri*.

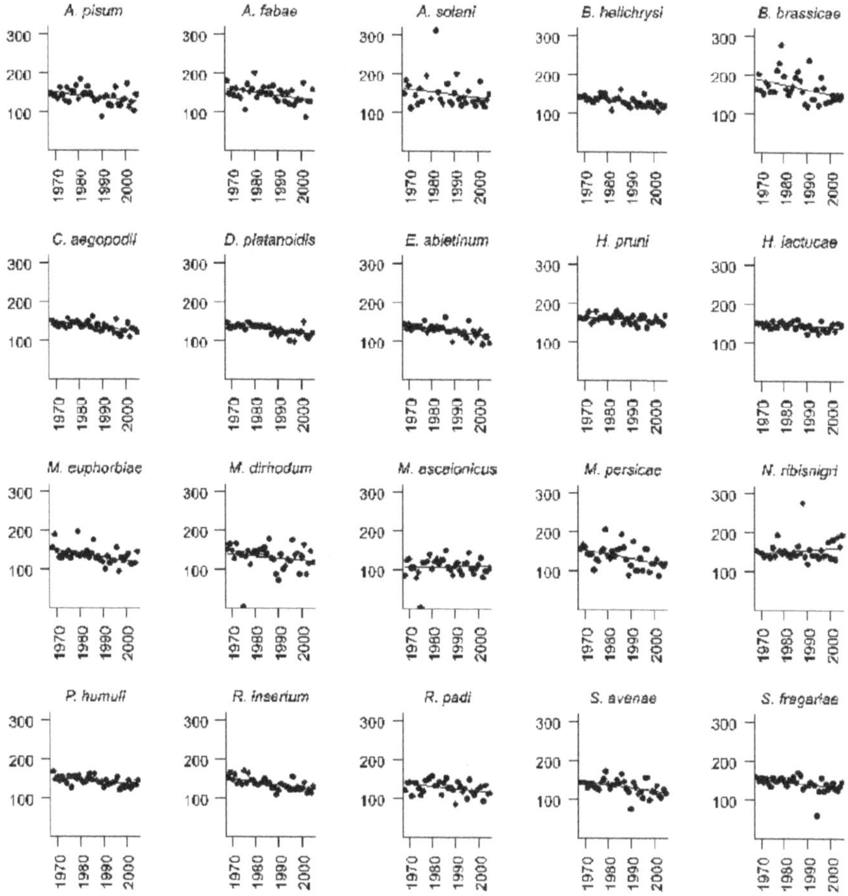

Fig. 3.5 Annual Julian date of first suction trap record (y-axis) against year (x-axis) for 20 aphid species at Rothamsted. Regression lines shown are from analyses before removing outliers for *A. solani*, *M. ascalonicus*, *S. fragariae*, *N. ribisnigri* and *M. dirhodum* (see text). Only lines for *A. solani*, *M. ascalonicus*, *N. ribisnigri* and *M. dirhodum* have a non-significant slope (P>0.05)

For *Myzus persicae*, date of last record advanced significantly (P<0.05) with year at one of 13 sites (Silwood, by 0.74 days per year) and became significantly later at none (Table 3.3, Fig. 3.8). Removal of the outliers for Dundee, East Craigs and Writtle had negligible effects on those analyses.

3.4 Discussion

Evidence has been found for the start of the flight season becoming earlier with warmer winters in 19 out of 20 species at Rothamsted and 12 out of 13 sites for *M. persicae*. In no case was there a significant retardation of the time of first record

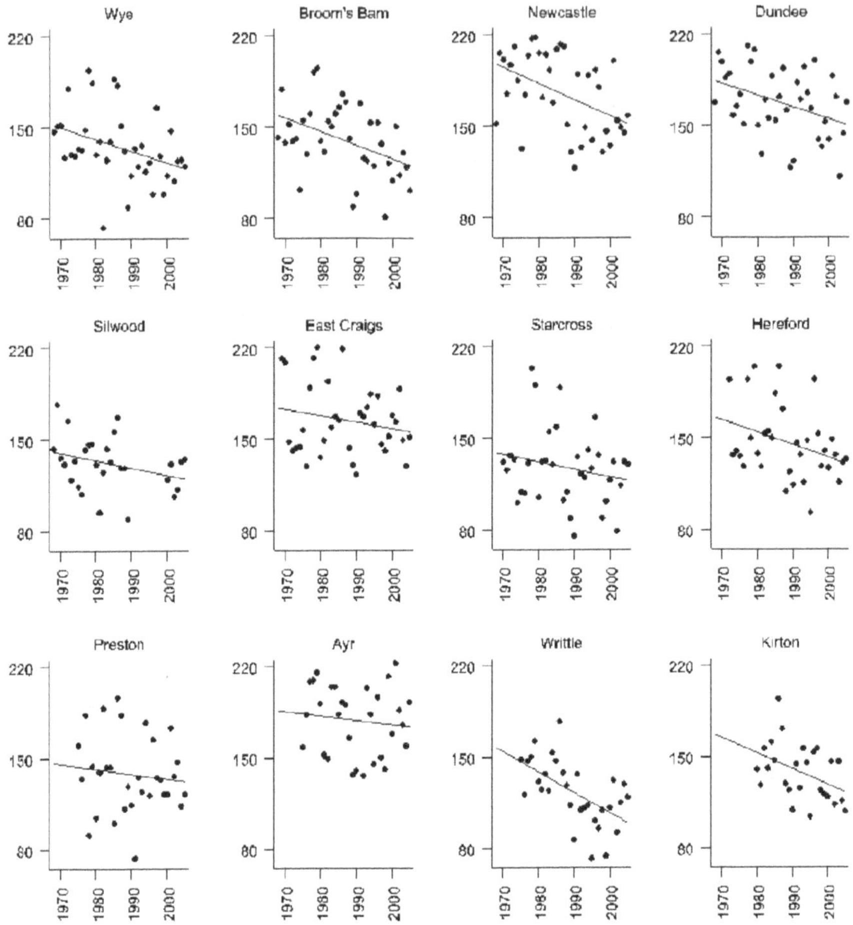

Fig. 3.6 Annual Julian date of first suction trap record (y-axis) against year (x-axis) for *Myzus persicae* at 12 suction trap sites with regression lines superimposed (Rothamsted shown in Fig. 3.5). Only lines for Ayr, East Craigs, Hereford, Preston, Silwood and Starcross have a non-significant slope (P>0.05)

with increasing winter temperature. There is also evidence for the relationships being stronger for those species that are continuously parthenogenetic at the site of interest. These effects have been shown previously (Harrington et al. 1990). The current analysis adds 17 more years' data and reveals more significant relationships for most species studied.

There is evidence for migrations starting earlier in later years in 16 out of 20 species at Rothamsted and seven out of 13 sites for *M. persicae*. In no case was there a significant retardation of the time of first record in later years. This is not surprising, bearing in mind that there has been a trend towards warmer winters in more recent years.

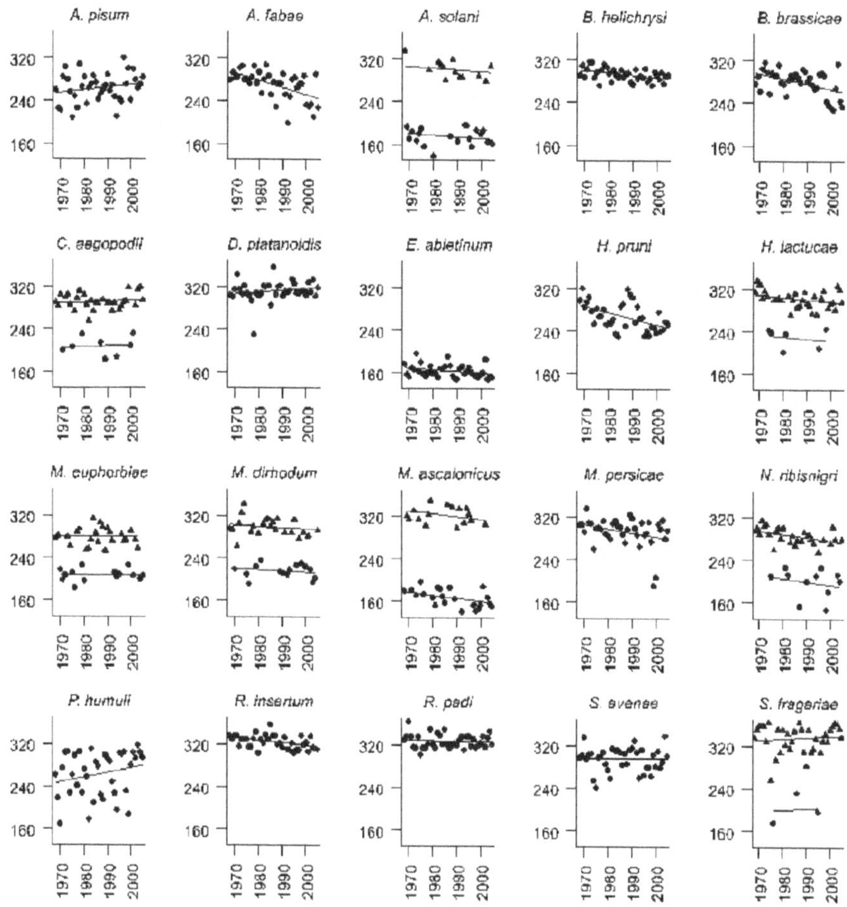

Fig. 3.7 Annual Julian date of last suction trap record (y-axis) against year (x-axis) for 20 aphid species at Rothamsted. Single regression lines shown are from analyses before removing outliers for *D. platanoidis* and *M. persicae* (see text). Of these only lines for *B. brassicae*, *A. fabae*, *H. pruni*, *R. insertum* and *B. helichrysi* have a significant slope (P<0.05). Parallel regression lines shown relate to groups of years with early (•) and late (▲) last records. Of these only lines for *M. ascalonicus* and *N. ribisnigri* have a significant slope (P<0.05)

There is no evidence that the end of the aphid flight period has become later in more recent years for any species. Indeed, the reverse is the case for seven species at Rothamsted and for *M. persicae* at Silwood. This at first seems surprising, as one might expect the flight season to be extended in warmer years because the threshold temperature for flight will be exceeded for longer. However, in the case of holocyclic species, night length is the major determinant of the winged aphids that return to winter hosts (Williams and Dixon 2007). It is thus interesting that five of the seven species for which there is a significant relationship are largely holocyclic and heteroecious (host alternating) at Rothamsted. A possible explanation for the advance

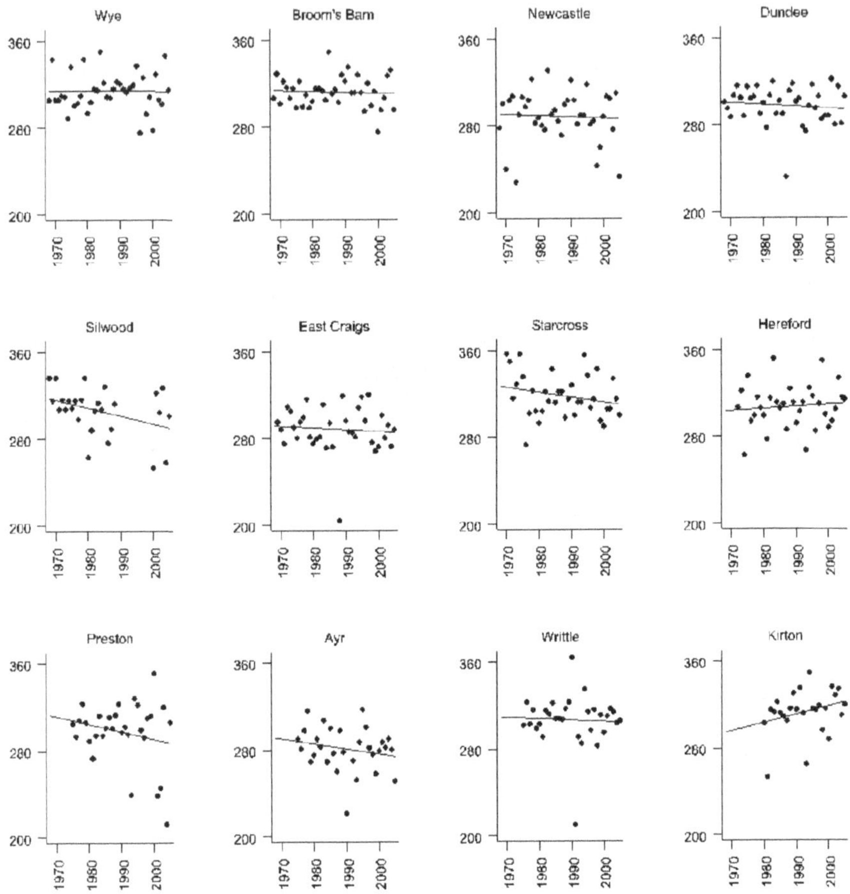

Fig. 3.8 Annual Julian date of last suction trap record (y-axis) against year (x-axis) for *Myzus persicae* at 12 suction trap sites (Rothamsted shown in Fig. 3.7). Regression lines shown are from analyses before removing outliers for Dundee, East Craigs and Writtle (see text). Only the regression line for Silwood has a significant slope ($P<0.05$)

in date of last flight with year is that, in warmer weather, the time from determination of winged forms by night length, to those winged forms becoming adult and flying, may be shortened. It would be interesting to analyse date of last suction trap record with respect to late summer and early autumn temperatures. The autumn migration of host alternating species must be optimised in relation to leaf fall on the primary host, allowing oviparae to complete their development and lay eggs before this occurs (Halkett et al. 2004). The relationship between timing of leaf fall and year is ambiguous (Menzel 2006). If leaf fall is unchanged with year, or is getting later, earlier migrations may be disadvantageous to host alternating aphids as they would benefit from a longer period on the secondary host, allowing further build up of numbers. If this is the case, it might be expected that, over time, selection will favour a longer critical night length for induction of autumn migrants, leading to later migrations. Another possible explanation for the advance in time of last aphid record is related to aphid abundance, and the fact that with lower abundance, the last aphid is likely to be detected earlier. Warm winters tend to lead to larger numbers of aphids early in the year (part of the reason for earlier detection). It has been shown that this can lead to lower numbers of aphids later in the year as a result of intrinsic density-dependent processes and increases in numbers of natural enemies (Blackman 1974; Dixon 1998). Further analyses are required to test whether this mechanism might explain the advance shown here in the date of the last suction trap record. The appearance of two distinct sets of points, early and late, for the last record of several species is intriguing. It may be that this is linked with the abundance of alatae in autumn, earlier ends to suction trap records being the result of the absence or small size of autumn migrations. Further analyses are required to test this.

There are many possibilities to extend these analyses. For example, it would be interesting to look at the relationships between winter temperature and the timing of first suction trap record for species that may be either continuously parthenogenetic, or holocyclic, along a gradient from exclusive adoption of one life cycle type to exclusive adoption of the other. It might be expected that the strength of the relationship will change systematically along this gradient. Data are available throughout Europe (Harrington et al. 2004), although the runs are longest for UK traps. In Poznan, Poland, which has very cold winters and where a suction trap has been operated since 1973, the relationship between winter temperature and date of first record of *M. persicae* is very weak but just significant ($P<0.05$) (Maria Ruszkowska, pers. comm.). Pan-European analysis of the timing of the first suction trap record in relation to a range of geographical, meteorological and land use variables have been attempted (Harrington et al. 2007). For *M. persicae*, the resulting model, which used ten explanatory variables, accounted for 54% of the variance in first suction trap record, similar to several of the sites in this study where mean January to February temperature was the only the explanatory variable tested.

Past studies have shown that the annual abundance of winged aphids in UK suction trap samples has tended not to change significantly with year (Harrington et al. 2003). However, numbers in spring and early summer have tended to increase, which is a component of the advancing date of time of first record. The planting or sowing of spring crops has tended not to advance so rapidly, as this depends particularly on soil moisture condition at planting, and not on winter temperature. Indeed, in the potato crop, date of 50% completion of planting in England has tended to become later since 1990 (Potato Council data). Aphids are thus tending to arrive earlier in the growing season when crops are more susceptible to feeding damage, and to the viruses transmitted, by aphids. For autumn-sown cereals, warmer winters increase the risk from *Barley yellow dwarf virus* (BYDV) because they lead to a greater prevalence of anholocycly and hence continued parthenogenesis on cereals, and because of the increased opportunity for movement of wingless, viruliferous aphids around the crop (Harrington 2003).

In general, climate change seems likely to be detrimental to UK growers with respect to damage resulting from aphids.

Acknowledgements We are grateful to all the UK suction trap operators and aphid identifiers, the Scottish Agricultural Science Agency for use of data from Ayr, Dundee and East Craigs, Maria Ruszkowska (Instytut Ochrony Roślin), and the Potato Council. The operation of the UK suction trap network is funded by the Lawes Agricultural Trust, the British Beet Research Organisation, the Scottish Executive Environment and Rural Affairs Department and by grant-aided support to Rothamsted from the Biotechnology and Biological Sciences Research Council (BBSRC) of the United Kingdom.

References

Blackman RL (1974) Aphids. Ginn & Company Limited, London

Dixon AFG (1998) Aphid ecology, 2nd edn. Chapman & Hall, London

Fleming RA, Tatchell GM (1995) Shifts in the flight periods of British aphids: a response to climate warming? In: Harrington R, Stork NE (eds) Insects in a changing environment. Academic Press, London, pp 505–508

Halkett F, Harrington R, Hullé M, Kindlmann P, Menu F, Rispe C, Plantegenest M (2004) Dynamics of production of sexual forms in aphids – theoretical and experimental evidence for adaptive 'coin-flipping' plasticity. Am Nat 63:E112–E125

Harrington R 2003. Turning up the heat on pests and diseases: a case study for Barley yellow dwarf virus. Proceedings of the British crop protection council international congress: CROP science and technology 2003, November 10th–12th 2003, Glasgow, UK, vol 2, pp 1195–1200

Harrington R, Clark SJ, Welham SJ, Verrier SJ, Denholm CH, Hullé M, Maurice D, Rounsevell MDA, Cocu N, EU EXAMINE Consortium (2007) Environmental change and the phenology of European aphids. Glob Chang Biol 13:1550–1564

Harrington R, Smith E, Hall M 2003. Assessing long-term trends in invertebrate biomass – a pilot study. Report to English Nature, June 2003

Harrington R, Tatchell GM, Bale JS (1990) Weather, life cycle strategy and spring populations of aphids. Acta Phytopathol Entomol Hung 25:423–432

Harrington R, Verrier P, Denholm C, Hullé M, Maurice D, Bell N, Knight J, Rounsevell M, Cocu N, Barbagallo S, Basky Z, Coceano P-G, Derron J, Katis N, Lukášová H, Marrkula I, Mohar J, Pickup J, Rolot J-L, Ruszkowska M, Schliephake E, Seco-Fernandez M-V, Sigvald R, Tsitsipis J, Ulber B (2004) 'EXAMINE' (EXploitation of Aphid Monitoring In Europe): an EU thematic network for the study of global change impacts on aphids. In: Simon J-C, Dedryver C-A, Rispe C, Hullé M (eds) Aphids in a new millennium. Proceedings 6th international aphid symposium, Rennes, France, 3–7 September 2001, pp 45–49

Harrington R, Woiwod I, 2007. Foresight from hindsight: the Rothamsted Insect Survey. Outlooks on pest management February 2007: 9–14

Leather SR, Walters KFA, Bale JS (1993) The ecology of insect overwintering. Cambridge University Press, Cambridge

Macaulay EDM, Tatchell GM, Taylor LR (1988) The Rothamsted Insect Survey '12-metre' suction trap. Bull Entomol Res 78:121–129

Menzel A (2006) Zeitliche Verschiebungen von Austrieb, Blüte, Fruchtreife und Blattverfärbung im Zuge der rezenten Klimaerwärmung. Forum für Wissen 2006:47–53

Sigvald R (1985) Mature plant resistance of potato plants against potato virus Yo (PVYO). Potato Res 27:285–290

Smith HG, Hallsworth PB (1990) The effect of yellowing viruses on yield of sugar beet in field trials, 1985 and 1987. Ann Appl Biol 116:503–511

Williams IS, Dixon AFG (2007) Life cycles and polymorphism. In: van Emden HF, Harrington R (eds) Aphids as crop pests. CABI, Wallingford, pp 69–85

Chapter 4
Consequences of Climate Change for Aphid-Based Multi-trophic Systems

Joan van Baaren, Cécile Le Lann, and Jacques JM van Alphen

Abstract Climatic models predict a 1.7–4.9°C increase in mean global temperatures from 1990 to 2100. In ecosystems in general, multitrophic interactions often result from a long co-evolutionary process specific to a particular environment and relatively stable climatic conditions. Temperature changes may differentially affect the biology of each of the component species of a system: for example, the herbivores, their natural enemies (parasitoids, predators and pathogens), and hyperparasitoids. The endosymbionts of these different insects are also affected, and their functions can be altered by temperature increase. Such effects could destabilise system dynamics even lead to extinctions. The effects of climatic change are likely to be relatively more important in higher trophic levels that depend on the capacity of lower trophic levels to adapt to these changes. This paper addresses the effects of climate on insect communities, focusing on aphids, aphids parasitoids and predators, and hyperparasitoids. For each trophic level, the general effect of temperature change on insects is discussed, with emphasis on species belonging to aphid-based communities. The effects of climate change on communities can be short-term or long-term. Short-term consequences include the direct effects of temperature on different life history traits such as development time (which affects the annual number of generations), metabolic rate (which affects activity levels, longevity, and fecundity), and sex allocation. Potential effects on endosymbiont survival, virus transmission, geographical distribution of species and phenological synchronisation between trophic levels are also discussed. Long-term effects involve genetic changes in populations associated with climatic adaptations.

Keywords Climate change · Parasitoids · Aphidiidae · Aphelinidae · Multitrophic interactions · Life history traits · Distribution · Phenology

---

J. van Baaren (✉)
UMR – CNRS 6553 ECOBIO, Université de Rennes I, 35 042, Rennes Cedex, France
e-mail: joan.van-baaren@univ-rennes1.fr

P. Kindlmann et al. (eds.), *Aphid Biodiversity under Environmental Change*,
DOI 10.1007/978-90-481-8601-3_4,

4.1 Introduction

Over the past 200 years, atmospheric CO_2 concentration has increased by 31% and other greenhouse gases and micron-size aerosols are now following the same trend and continue to rise. The atmoshopheric concentration of CO_2 has a direct effect on plant growth, and by extension, on phytophagous insects (see Chapter 2 by Ameixa, in this volume). This increase in CO_2 concentration has been implicated as one cause of the increase in average global temperature. Based on projected increases of these greenhouse gases, climatic models predict a 1.7–4.9°C global warming from 1990 to 2100 (Karl and Trenbeth 2003). This increase in temperature will have a greater effect on insects than the rising CO_2 concentration (Bezemer et al. 1998; Harrington et al. 2001). All trophic levels stand to be affected: the herbivores, their natural enemies (parasitoids, predators and pathogens), and hyperparasitoids and tertiary predators. The endosymbionts of these different insects will also be affected, and their functions may be altered by increasing temperature. The impact of climate change is likely to be relatively more important in higher trophic levels that depend on the capacity of the lower trophic levels to adapt to these changes. Parasitoids, predators and hyperparasitoids are thus organisms for which severe effects are expected, as they represent the third and fourth trophic levels (Hance et al. 2007).

In ecosystems, multitrophic interactions are the result of a long co-evolutionary process that unfolds within a particular environment and against a background of specific climatic conditions. Temperature changes may differentially affect the biology of each component species of a system (Fig. 4.1), potentially destabilising the dynamics and leading to the extinction of part of the system (van der Putten et al. 2004). Parasitism by parasitoids and hyperparasitoids, and predation, depends both on the capacity of adult insects to locate and select their hosts or prey and, for the parasitoids, on the capacity of the larvae to evade or overcome the host's immune response and subvert, or regulate, the host's physiology. Many of these interactions

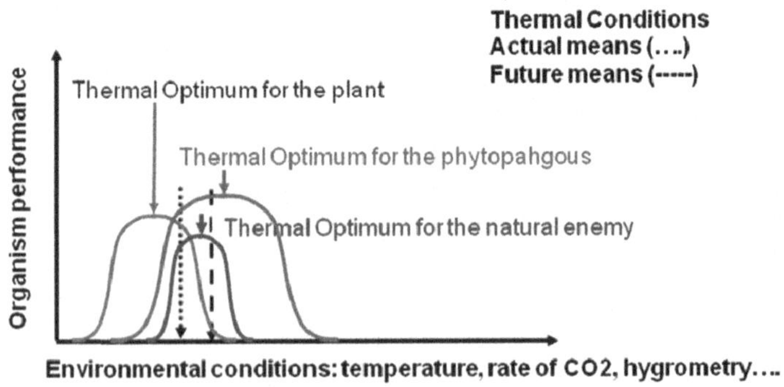

Fig. 4.1 Thermal optimum for each component of a tri-trophic system. In this example, the thermal conditions are more favourable to the plant than the herbivore or natural enemy, but an increase of the mean temperature could change the outcome

are temperature-dependent and could be disrupted by changes in environmental conditions (Hance et al. 2007).

This paper addresses the general effect of temperature change on aphids, aphid parasitoids and predators, and hyperparasitoids. Few studies have shown effects of CO_2 on the third or higher trophic levels, probably because the effects of rising CO_2 across three trophic levels are very complex (Chen et al. 2007). In the rare studies on effects of CO_2, the effects of temperature are rarely considered.

Aphids are herbivorous insects that are often abundant and can be found in most terrestrial habitats; in temperate regions, one plant species in four is subject to aphid attack (Dixon 1998). Aphids are attacked by solitary hymenopterous parasitoids belonging to the Aphidiinae (Ichneumonoidea, Braconidae) and Aphelinidae (Chalcidoidea) (Van Veen et al. 2008). Primary aphid parasitoids are attacked by two groups of hyperparasitoids. Members of the first group lay eggs inside the bodies of primary parasitoid larvae within the aphid before mummification; nearly all these belong to the Alloxystiniae (Figitidae, Charipinae). Members of the second group attack the mummy stage, irrespective of whether it contains primary or secondary parasitoids. They paralyze the mummy inhabitant and their larvae develop immediately (Traugott et al. 2008). This group consists of several unrelated genera of Pteromalidae and a single genus (*Dendrocerus*) of Megaspilidae.

The effects of climate change on communities can be both short-term and long-term. Short-term consequences include the direct effects of temperature on different life history traits; development time (which affects the annual number of generations), metabolic rate (which affects activity levels, longevity, and fecundity), and sex allocation. Endosymbiont survival, virus transmission, host distribution and phenological synchronisation between trophic levels also will be affected. Long-term effects involve genetic changes in populations associated with climatic adaptations, but since such studies are scarce and have not addressed aphid-based communities, discussion of these will be brief.

4.2 Life History Traits

Most insect life history traits are linked to temperature and numerous studies have shown that temperature affects one or more traits in each species. However, species interact with one another and, to evaluate temperature effects on these interactions, similar studies need to be performed on all the species in a community, something, which has never been done.

4.2.1 Development Time and Population Dynamics

For most insect populations, the intrinsic rate of increase gradually increases with temperature until some threshold, after which it decreases abruptly. The increase phase is generally linked to a decrease in development time and an increase in reproductive rate. For example, the development time of the aphid *Aphis gossypii* Glover (Homoptera: Aphididae), an important pest in glasshouse crops, ranged from 4.8 days at 20°C to 3.2 days at 30°C, but the intrinsic rate of increase was greatest at 25°C ($r_m = 0.556$ day^{-1}). At 20 and 30°C, the intrinsic rate of increase was 0.426 and 0.510, respectively (Vansteenis and Elkhawass 1995). At 25°C, an average of 12 days is needed for the parasitoid *Aphidius matricariae* (Haliday) (Homoptera: Aphididae) to complete its development from oviposition to adult emergence on this host, and its development is shortest at this temperature (Zamani et al. 2007). In such systems, the aphid generation time is shorter than that of parasitoids. For both species, temperatures above 25°C induce a decrease in performance.

In some systems, thermal optima are not the same for interacting species. For example, the parasitoid *Aphidius colemani*, which can also attack *Aphis gossypii*, performs better at 30°C. Thus, the consequences of increasing temperature for population dynamics of parasitoid-host systems depend on whether the host or the parasitoid is more affected. In aphid parasitoid or predator systems, the aphids always have a shorter generation time than their natural enemies, but temperature can mediate the magnitude of the disparity. In addition, temperature affects other population parameters, e.g., fecundity, longevity, and parasitoid or predator attack rate. For example, below 11°C, the reproductive rate of the pea aphid, *Acyrthosiphon pisum* (Harris) (Homoptera: Aphididae) exceeds the rate at which the coccinellid *Coccinella septempunctata* L. can consume it, but above 11°C, the reverse occurs (Harrington et al. 2001).

4.2.2 Metabolic Rate, Activity, Longevity and Fecundity

It is expected that an increase in temperature will result in an increase in metabolic rate, and as a consequence, the general activity of an insect. For natural enemies, this could result in an increase in fecundity if it results in a higher rate of attack on prey. However, it will also result in a decrease in longevity due to the accelerated use of energy. The mean and maximum walking speeds of the parasitoid *Aphelinus asychis* Walker (Hymenoptera: Aphelinidae) increase with increase in temperature from 8.5 to 28.5°C (Mason and Hopper 1997) and are associated with an increase in the rate of parasitism. Parasitism rates of *A. colemani* and *A. matricariae*, estimated as the percentage of aphids mummified, increase almost linearly with temperature to a maximum at 25°C (Zamani et al. 2007). However, the aphid birth rate also increases: *A. gossypii* produces 65.9 and 69.8 nymphs/female, respectively, at 25 and 30°C, compared at 59.9 nymphs/female 20°C (Vansteenis and Elkhawass 1995). Immature mortality was approximately 20% and did not differ among temperatures, likely because most mortality occurred during the first instar after only a short exposure time.

An increase in metabolic rate is accompanied by a decrease in longevity in all insects. Female life span in the aphid hyperparasitoid *Asaphes vulgaris* Walker

(Hymenoptera: Pteromalidae) reared on the primary parasitoid *Aphidius nigripes* Ashmead (Hymenoptera: Aphidiidae) is about 4 months at 15°C, compared with 46 days at 25°C. Under the same conditions, male longevity is 66 and 19 days, respectively (Brodeur and McNeil 1994).

Aphid parasitoids must actively search for food and hosts, whereas aphids are generally fixed on their host plant. However, the defensive behaviour of aphids is also affected by temperature: when temperature increases, pea aphids show a reduced tendency to drop from the plant in response to the presence of natural enemies. This might decrease their risk of dehydration while off the plant, but renders them more vulnerable to parasitoids (Stacey and Fellowes 2002).

4.2.3 Sex Allocation

Temperature may influence sex allocation in arrhenotokous hymenopteran parasitoids, although the effects vary among species. The sex ratio of *Aphelinus varipes* Forster (Hymenoptera: Aphelinidae) reared on *A. gossypii* is female-biased and varied from 92% female when reared at 25°C, to 70% at 20°C but is not significantly influenced by host stage (Rohne 2002). However, when the parasitoid *Diaeretiella rapae* M'Intosh (Hymenoptera: Aphidiidae) was reared on the host *Diuraphis noxia* (Kurdjumov) (Homoptera: Aphididae), the percentage of females was lowest at 26.7°C but did not vary between 10 and 21°C (Bernal and Gonzalez 1997).

The effect of temperature on sex allocation is complicated by the fact that, in several parasitoid species, sex ratio is influenced by host size, with males more likely to emerge from smaller hosts. Host size may also be linked to temperature, as in the black bean aphid, *Aphis fabae* Scopoli, which is larger when reared at low temperatures than at high (Li and Mills 2004).

4.2.4 Effects of CO_2 Increase on Life History Traits

In general, enhanced CO_2 levels increase photosynthesis, growth, yield and C:N ratios in most plant species, particularly C3 plants (Pritchard et al. 1999) which may affect the quality and quantity of food available to insect herbivores (Chen et al. 2007). Using a tri-trophic system comprised of wheat, *Triticum aestivum* L., *Sitobion avenae* (Homoptera: Aphididae) and the predator *Harmonia axyridis* Pallas (Coleoptera: Coccinellidae) and the parasitoid *Aphidius picipes* Nees (Hymenoptera: Aphidiidae), Chen et al. (2007) show that elevated levels of CO_2 (550 and 750 ml/l vs. ambient CO_2) have little effect on the growth and development of the predator, but a negative effect on these traits in the parasitoid. However, both the predator and the parasitoid increased their rate of feeding/parasitism at high levels of CO_2 (Chen et al. 2007). Chen et al. (2005) found that the larvae of *H. axyridis* consume more cotton aphids, *A. gossypii,* reared on cotton plants grown at elevated CO_2 levels and so compensated for the reduced soluble protein levels in the aphids that resulted from the decrease in foliar N and increase in C:N ratio. Under these conditions, significantly more aphids reared on plants grown at elevated CO_2 levels were consumed by lady beetle larvae than those from plants grown at ambient CO_2 levels.

A study was conducted by Stacey and Fellowes (2002) on the affect of elevated CO_2 on species interactions across three trophic levels: a plant (*Brassica oleracea*), two aphid herbivores (the generalist *Myzus persicae* Sulzer and the specialist *Brevicoryne brassicae* L.), and two natural enemies, the predator *Hippodamia convergens* (Coleoptera: Coccinellidae) and the parasitoid *Diaeretiella rapae* (Hymenoptera: Aphidiidae). Elevated CO_2 produced plants that were larger and had decreased water and nitrogen content. *Brevicoryne brassicae* reared on such plants were larger and accumulated more fat, while *M. persicae* was unaffected. Increasing CO_2 concentrations altered the outcome of competition between the two aphid species, to the advantage of *B. brassicae*. Changes in plant quality resulting from the CO_2 treatments did not alter the number of aphids consumed by *H. convergens*, or parasitised by *D. rapae*, suggesting that the quality of the aphids as prey/hosts did not change.

In another study of the effects of elevated CO_2 (ambient + 200 μmol/mol) on a tri-trophic system consisting of a plant, *B. oleracea*, an aphid, *M. persicae*, and a parasitoid, *A. matricariae*, the total above-ground plant biomass, foliar nitrogen and carbon concentration at the end of the experiment were unaffected, but aphid abundance was enhanced (Bezemer et al. 1998). Parasitism rates remained unchanged at elevated levels of CO_2, suggesting that *M. persicae* might become more problematic in a climate change scenario (Bezemer et al. 1998).

The inconsistent findings of these three studies suggest that the effects of elevated CO_2 on aphid communities depend on the particular species involved and that it is not possible to generalize.

4.3 Endosymbionts

Parasitoids and their hosts may harbour endosymbiotic bacteria that can influence various aspects of their life history. Aphids have a variety of bacterial symbionts including an obligate association with the mutualist *Buchnera aphidicola* (Munson et al. 1991; Shigenobu et al. 2000). In addition, they may possess secondary symbionts, facultative associations with several heritable bacteria that may be conditionally beneficial or deleterious (Degnanj and Moran 2008). These secondary symbionts belong to *Enterobacteriaceae* and are known from the pea aphid, *A. pisum*, as well as from various other species and subfamilies of aphids, with few records from other insect hosts. The symbionts can be transferred among hosts in the laboratory and aphids can be freed of their endosymbionts with antibiotics, allowing experimental studies of their effects on hosts (Moran et al. 2005). These studies have reveal that these bacteria have major effects on aphid biology, including resistance to parasitoid wasps, tolerance to heat stress, restoration of reproduction in aphids experimentally deprived of *B. aphidicola*, and changes in host plant range (Tsuchida et al. 2004; Moran et al. 2005). Both major groups of endosymbionts, *Wolbachia* and *Buchnera*, may be negatively affected or even eliminated by short exposures to high temperature (Thomas and Blandford 2003).

In field populations of *A. pisum*, the non-obligate endosymbiont *Hamiltonia defensa* occurs at intermediate frequencies and includes strains providing variable levels of host protection against parasitoids (Oliver et al. 2005). Thus, the acquisition and loss of particular *H. defensa* strains is likely to have a dramatic influence on aphid fitness in nature. The effect of secondary endosymbionts varies with temperature. Cheng et al. (2000) showed that, the presence of these endosymbionts protect the aphid *A. pisum* against heat stress at 25°C, whereas they decrease fecundity at 20°C. The defensive immunity conferred to aphids by *H. defensa* fails under heat stress (Bensadia et al. 2006). Similarly, *B. aphidicola* produces an amino acid essential for the development of the aphid parasitoid *Aphidius ervi* (Rahbé et al. 2002) and is sensitive to high temperatures, which may eliminate the endosymbiont (Ohtaka and Ishikawa 1991). In conclusion, although definitive studies have not been done, climate change is expected to have broad-ranging impacts on aphid-based communities via effects on their associated endosymbionts.

4.4 Virus Transmission by Aphids

Barley yellow dwarf virus (BYDV) is one of the world's most severe viral diseases of autumn-sown cereals and is caused by members of the family Luteoviridae. BYDV is transmitted in a persistent, circulative, non-propagative manner by several aphid species. Yield losses of barley and wheat are mainly due to autumn infections with PAV (*Padi avenae* virus), which is transmitted in 90% of cases by the aphid *Rhopalosiphum padi* (L.) (Homoptera: Aphididae). Environmental conditions govern the pattern and timing of primary infections by viruliferous alates and the speed of subsequent virus dissemination within a field by apterae (secondary spread). Temperature influences both the population dynamics of the vector (5°C is the developmental threshold for *R. padi*), and the virus transmission process (15°C is the take-off threshold for alatae). In autumn and winter, some *R. padi* genotypes enter a phase of sexual reproduction and produce males and gynoparae, whereas other *R. padi* genotypes continue to reproduce parthenogenetically (Simon et al. 2002) and continue to spread BYDV. The warmer the January to August period, the higher the percentage of viruliferous aphids the following autumn, perhaps due to an increased rate of population growth and, consequently, the rate of virus transmission in agroecosystems (Fabre et al. 2005). Similarly, Foster et al. (2004) showed that spring incidence of virus in the UK from 1995 to 1998 was related to the incidence of aphids the preceding autumn. Both virus and aphid incidences are likely alsobe related to a range of crop and field characteristics such as sowing date, geographical location, topography and climate, the proximity of the field to the sea, the extent of arable land in the vicinity of the field, and the aspect and size of the field. Thus, as temperature is a major factor influencing virus transmission via effects on aphid biology, climate change is expected to modify the incidence of aphid-borne virus infections.

4.5 Changes in the Geographical Distribution of Hosts and Parasitoids

The geographical distribution of plants, herbivores, and parasitoids may change as a result of various processes, including changes in the dispersal capacity of insects and their host plants, the appearance of green bridges (new plant species providing favourable overwintering sites), and the destabilization of resident ecosystems (Cannon 1998; Parmesan 2006). Within an area where abiotic conditions are tolerable for an herbivore, its distribution may be limited by the availability of its host plant. Likewise the distribution of a parasitoid may be limited by the availability of their hosts. Therefore, parasitoids will be more often constrained within their range than are their hosts.

A few examples are known of aphid-parasitoids colonizing new regions. In the Spitzbergen archipelago, Hullé et al. (2008) found two endemic aphid species, *Acyrthosiphon svalbardicum* Heikinheimo and *Sitobion calvulum* Ossiannilsson. Both were parasitized by *D. rapae*, which must have colonized the archipelago from some unknown southern source utilizing endemic hosts already present. Seven species of aphids have colonized this archipelago, but without surviving more than a few months (Hulle, pers. comm.). Unfavourable abiotic conditions are most likely responsible for failures to establish. Three species of aphids were introduced onto Marion Island, *R. padi*, *Myzus ascalonicus* Doncaster and *Macrosiphum euphorbiae* Thomas (CraVord et al. 1986). A single parasitoid species, *A. matricariae* (Haliday), became established as early as 2001, likely as a consequence of accidental introduction. *Aphidius matricariae* is a cosmopolitan species known from at least 81 countries, including the nearby South African mainland (Yu et al. 2005). Approximately 115 aphid species are parasitized by *A. matricariae* (Lee et al. 2007). On Marion Island, *A. matricariae* seems to be restricted to a single host species (*R. padi*), despite the presence of two other aphid species known to be utilized as hosts elsewhere (Gillespie et al. 2002; Yu et al. 2005; Lee et al. 2007).

Differences in thermal tolerances among hosts and their parasitoids may result in differences in their respective geographic distributions. Three closely related aphid parasitoids co-occur in cereal crops in Brittany (France): *Aphidius ervi* Haliday, *Aphidius avenae* Haliday, and *Aphidius rhopalosiphi* De Stefani Perez. All attack the grain aphid, *Sitobion avenae*, but differ in thermal tolerance and are less resistant than their host to high temperatures (Fig. 4.2).

Phytophagous species often extend their geographic distribution northward in temperate regions following an increase in temperature (Parmesan 2006 for a review). These herbivores could damage agricultural crops or forests if they are

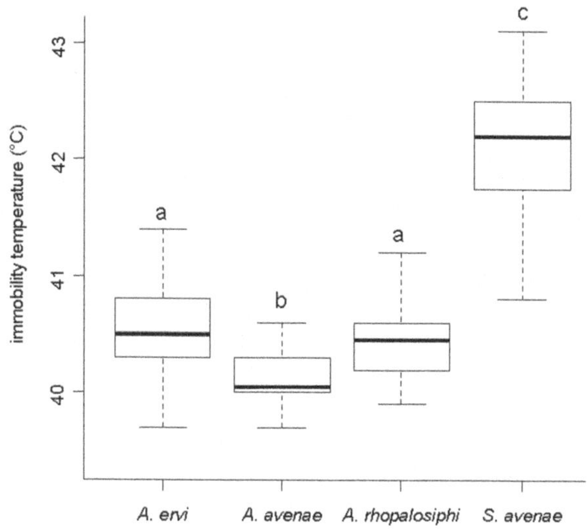

Fig. 4.2 Critical high temperatures of females of three species of parasitoids: *A. ervi*, *A. avenae*, *A. rhopalosiphi* and their preferred second instars hosts of *S. avenae*. Critical high temperatures were obtained by measuring the temperature at which individuals did not move anymore when subjected to an increase in temperature of 1°C per min. These critical temperatures are thought to be ecologically relevant and a useful way of assessing the range of activity of populations (Terblanche et al. 2007). The protocol used was adapted from Powell and Bale (2004). GLM F test =118.04; $p < 0.0001$. Different *letters* indicate significant differences between species. Median and maximum and minimum values are shown

not followed by their natural enemies. In aphid-based communities, both parasitized and unparasitized aphids can be dispersed by wind (Hulle, pers. comm.). Although dispersal of parasitized aphids could result in the dispersal of parasitoids, if hosts are better at dispersing than their parasitoids, the latter may be less able to escape regions that have become unfavourable, which, in combination with lower resistance to high temperatures, could result in local extinctions of parasitoid populations.

4.6 Phenological Synchronisation and Diapause

There are few empirical studies (but numerous models) on the effect of primary parasitoids on aphid population dynamics, or that of hyperparasitoids on primary parasitoid populations (and via these on aphid populations). Most such studies address aphid parasitoid systems in cereal crops. As many as 33% of cereal aphids may be parasitized (Traugott et al. 2008). Aphid primary parasitism tends to increase monotonically with aphid population growth. As in previous studies (Walton et al. 1990; Sullivan and Völkl 1999), Traugott et al. (2008) found that rates of hyperparasitism increase towards the end of the period of aphid activity, although overall rates were low. In contrast, Höller et al. (1993) found that rates of primary parasitism declined late in the season. The authors hypothesized that primary females leave the field as hyperparasitoid densities increase, resulting in a drop in primary parasitism rates in the latter part of the season, possibly because they avoid areas contaminated with chemical residues of hyperparasitoids.

Among the conditions thought to maintain stability in host-parasitoid systems are a duration of parasitoid development between 0.5 and 1.5 times that of its host and a juvenile development time of both host and parasitoid that is longer than the lifespan of the adult (Godfray et al. 1994). Any change in these ratios would affect the dynamics of the host-parasitoid interaction. The phenological synchrony between plants and insects, and between hosts and parasitoids, could become uncoupled if the two processes are temperature-driven in different ways. There are few studies of the second and third trophic levels, but the different indices developed below suggest that their synchrony could be affected. Preliminary results of flexible, age-structured models of host-parasitoid interactions suggest that annual variation in phenological asynchrony may be an important destabilizing factor. It is necessary to consider rare weather events as well as long-term averages. For example, if a parasitoid species has a lower base temperature than its host, in a warm spring the parasitoid could emerge earlier in the seasonal development of its host. When these conditions occur at a frequency of once in 25 years or less often, the system remains stable and persists, whereas a frequency above 1 in 20 can drive the system to local extinction (Godfray et al. 1994). When parasitoids emerge early, the host population is exposed to a relatively large number of searching adult parasitoids while it is still small and is considerably reduced as a consequence. The few hosts that survive produce only a small generation the following year and, causing the parasitoid population to crash to low levels, and allowing the host population to rapidly increase (Godfray et al. 1994).

Similarly, in a model of an aphid-parasitoid system, Hoover and Newman (2004) showed that the time of parasitoid arrival in an aphid population influences the growth of the parasitoid population and its impact on the host population. If the parasitoid arrives either too early (before the aphid population has entered the exponential growth phase) or too late, negligible parasitoid offspring are produced. When it arrives too early, either the parasitoid population disappears before the aphid population increases or the parasitoid eliminates the small aphid population before it reaches the exponential growth phase. In contrast, if the parasitoid arrives during the exponential growth phase, large parasitoid populations will result. The model of Hoover and Newman (2004) also shows that the responses of aphids to environmental changes are the same in the presence or absence of parasitoids, suggesting that there may be little or no interaction between parasitism and environmental changes as far as the aphids are concerned. These results suggest that even small climatic changes that affect the synchrony of parasitoid activity with host populations could have a large effect on the population dynamics of both. However, this model simulates long-term global changes (i.e., an increase of the mean temperature), not the impact of transient episodes of extreme temperatures.

Several models (Newman 2005) have shown that the abundance of aphids in cereal fields in the spring in Western Europe can be largely predicted by the temperatures in February. Hansen (1999) has shown that it is possible to relate the arrival of *R. padi* in cereals in Denmark after 1 May to day-degrees in February, and to temperature and precipitation during the following months.

In oceanic regions of western Europe, parasitoids attacking cereal aphids are able to survive for periods without their hosts. In summer, arrested development of *A. rhopalosiphi* permits this important parasitoid of cereal aphids to extend its development period up to 4 months. This arrested development is fairly infrequent but ensures the survival of a part of the population. A longer duration of summer development is recorded for a few hyperparasitoids (Krespi et al. 1997). In winter in Belgium, only a fraction of the parasitoid population associated with anholocyclic *S. avenae* enter diapause. Part of the population undergoes quiescence and may become active during winter if climatic conditions allow them to terminate their development and emerge. The plasticity of diapause induction may allow an efficient exploitation of available resources and, consequently, non-diapausing parasitoids should only be associated with anholocyclic aphids that remain reproductive throughout winter. In contrast, parasitoids associated with holocyclic aphids should enter diapause during winter because no hosts are available. The presence of both diapausing and non-diapausing mummies in parasitoid species attacking anholocyclic *S. avenae* may lower the risk of their extinction in the face of unpredictible winter weather. To survive winter, at least a part of the parasitoid populations must be able to tolerate the annual minimum temperature in their habitat.

4.7 Conclusions

Species may adapt to long-term changes in climate if these changes do not occur too abruptly. Short generation times and large effective population sizes may facilitate such adaptations. In systems where hosts have shorter generation times than their parasitoids, hosts may adapt to climate change faster than their parasitoids.

The effects of climatic change cannot be generalized across different host-parasitoid systems. For example, the three major parasitoids of *S. avenae* in cereal fields in Denmark have different lower temperature thresholds and different degree-day requirements for development. As temperature increases, the rate of superparasitism by *A. ervi* increases and the proportion of males increases in *Praon volucre* (Haliday), whereas these factors do not change in the other species (Sigsgaard 2000). Although it is certain that changes in these systems will occur as a result of climate change, the precise outcomes are difficult to predict. However, most models of host-parasitoid interactions predict an increase in pest-outbreaks, due to increases in the annual number of host generations (Landsberg and Smith 1992; Cannon 1998; Bezemer et al. 1998; Harrington et al. 2001).

Most of the cited studies focus on the effects of a temperature increase on different traits or interactions in host-parasitoid systems. However, an increase in temperature is not the only consequence of climate change and more frequent heat waves, droughts and extreme precipitation events can be expected (Easterling et al. 2000). For most organisms, including insects, the unpredictable occurrence and higher frequency of extreme environmental conditions may present greater challenges than the gradual increase in average temperature (Godfray et al. 1994).

References

Bensadia F, Boudreault S, Guaya F, Michaud D, Cloutier C (2006) Aphid clonal resistance to a parasitoid fails under heat stress. J Insect Physiol 52:146–157

Bernal J, Gonzalez D (1997) Reproduction of *Diaeretiella rapae* on Russian wheat aphid hosts at different temperatures. Entomol Exp Appl 82:159–166

Bezemer MT, Jones TH, Knight KJ (1998) Long-term effects of elevated CO_2 and temperature on populations of the peach potato aphid *Myzus persicae* and its parasitoid *Aphidius matricariae*. Oecologia 116:128–135

Brodeur J, Mc Neil JN (1994) Life-history of the aphid hyperparasitoid *Asaphes vulgaris* Walker (Pteromalidae) – Possible consequences on the efficacy of the primary parasitoid *Aphidius nigripes* Ashmead (Aphidiidae). Can Entomol 126:1493–1497

Cannon RJC (1998) The implications of predicted climate change for insect pests in the UK, with emphasis on nonindigenous species. Glob Chang Biol 4:785–796

Chen FJ, Ge F, Parajulee MN (2005) Impact of elevated CO_2 on tri-trophic interaction of *Gossypium hirsutum*, *Aphis gossypii*, and *Leis axyridis*. Environ Entomol 34:37–46

Chen FJ, Wu G, Megha N, Parajulee MN, Ge F (2007) Impact of elevated CO_2 on the third trophic level: a predator *Harmonia axyridis* and a parasitoid *Aphidius picipes*. Biocontrol Sci Technol 17:313–324

Cheng DQ, Montllor CB, Purcell AH (2000) Fitness effects of two facultative endosymbiotic bacteria on the pea aphid, *Acyrthosiphon pisum*, and the blue alfalfa aphid, *A. kondoi*. Entomol Exp Appl 95:315–323

CraVord JE, Scholtz CH, Chown SL (1986) The insects of sub-Antarctic Marion and Prince Edward Islands; with a bibliography of entomology of the Kerguelen Biogeographical Province. S Afr J Antarct Res 16:42–84

Degnanj PH, Moran NA (2008) Evolutionary genetics of a defensive facultative symbiont of insects: exchange of toxin-encoding bacteriophage. Mol Ecol 17:916–929

Dixon AFG (1998) Aphid ecology. Chapman and Hall, London

Easterling DR, Meehl GA, Parmesan C, Changnon SA, Karl TR, Mearns LO (2000) Climate extremes: observations, modeling, and impacts. Science 289:2068–2074

Fabre F, Plantegenest M, Mieuzet L, Dedryver C, Leterrier JL, Jacquot E (2005) Effects of climate and land use on the occurrence of viruliferous aphids and the epidemiology of barley yellow dwarf disease. Agric Ecosyst Environ 106:49–55

Foster GN, Blake S, Tones SJ (2004) Occurrence of barley yellow dwarf virus in autumn-sown cereal crops in the United Kingdom in relation to field characteristics. Pest Manag Sci 60:113–125

Gillespie DR, Shipp JL, Raworth DA, Foottit RG (2002) *Aphis gossypii* Glover, melon/cotton aphid, *Aulacorthum solani* (Kaltenbach), foxglove aphid, *Macrosiphum euphorbiae* (Thomas), potato aphid, and *Myzus persicae* (Sulzer), green peach aphid (Homoptera: Aphididae). In: Mason PG, Huber JT (eds) Biological control programmes in Canada, 1981–2000. CAB, Oxon, pp 44–49

Godfray HCJ, Hassell MP, Holt RD (1994) The population dynamic consequences of phenological asynchrony between parasitoids and their hosts. J Anim Ecol 63:1–10

Hance T, van Baaren J, Vernon P, Boivin G (2007) Impact of temperature extremes on parasitoids in a climate change perspective. Ann Rev Entomol 52:107–126

Hansen LM (1999) Effect of weather during spring on the time of arrival of bird cherry-oat aphid (*Rhopalosiphum padi* L.) in spring barley (*Hordeum vulgare* L.) fields. Acta Agric Scandinavia B 49:117–121

Harrington R, Fleming RA, Woiwod IP (2001) Climate change impacts on insect management and conservation in temperate regions: can they be predicted? Agric For Meteorol 3:233–240

Höller C, Bargen H, Vinson SB, Braune HJ (1993) Sources of the marking pheromones used for host discrimination in the hyperparasitoid *Dendrocerus carpenteri*. J Insect Physiol 39:649–656

Hoover JK, Newman JA (2004) Tritrophic interactions in the context of climate change: a model of grasses, cereal aphids and their parasitoids. Glob Chang Biol 10:1197–1208

Hullé M, Bonhomme J, Maurice D, Simon JC (2008) Is the life cycle of high arctic aphids adapted to climate change? Polar Biol 31:1037–1042

Karl TR, Trenbeth KE (2003) Modern global climate change. Science 302:1719–1723

Krespi L, Dedryver C, Creach V (1997) Variability in the development of cereal aphid parasitoids and hyperparasitoids in oceanic regions as a response to climate and abundance of hosts. Environ Entomol 26:545–551

Landsberg J, Smith MS (1992) A functional scheme for predicting the outbreak potential of herbivorous insects under global atmospheric change. Aust J Bot 40:565–577

Lee JE, Slabber S, van Vuuren BJ, van Noort S, Chown SL (2007) Colonisation of sub-Antarctic Marion Island by a non-indigenous aphid parasitoid *Aphidius matricariae* (Hymenoptera, Braconidae). Polar Biol 30:1195–1201

Li BP, Mills N (2004) The influence of temperature on size as an indicator of host quality for the development of a solitary koinobiont parasitoid. Entomol Exp Appl 110:249–256

Mason PG, Hopper KR (1997) Temperature dependence in locomotion of the parasitoid *Aphelinus asychis* (Hymenoptera: Aphelinidae) from geographical regions with different climates. Environ Entomol 26:1416–1423

Moran NA, Degnan PH, Santos SR, Dunbar HE, Ochman H (2005) The players in a mutualistic symbiosis: insects, bacteria, viruses, and virulence genes. Proc Natl Acad Sci USA 102:16919–16926

Munson MA, Baumann P, Clark MA (1991) Evidence for the establishment of aphid-eubacterium endosymbiosis in an ancestor of four aphid families. J Bacteriol 173:6321–6324

Newman JA (2005) Climate change and the fate of cereal aphids in Southern Britain. Glob Chang Biol 11:940–944

Ohtaka C, Ishikawa H (1991) Effects of heat treatment on the symbiotic system of an aphid mycetocyte. Symbiosis 11:19–30

Oliver KM, Moran NA, Hunter MS (2005) Variation in resistance to parasitism in aphids is due to symbionts not host type. Proc Natl Acad Sci USA 102:12795–12800

Parmesan C (2006) Ecological and evolutionary responses to recent climate change. Ann Rev Ecol Evol Syst 37:637–669

Powell SJ, Bale JS (2004) Cold shock injury and ecological costs of rapid cold hardening in the grain aphid *Sitobion avenae* (Hemiptera: Aphididae). J Insect Physiol 50:277–284

Pritchard SG, Rogers HH, Prior SA, Peterson CM (1999) Elevated CO_2 and plant structure: a review. Glob Chang Biol 5:807–837

Rahbé Y, Digilio MC, Febvay G, Guillaud J, Fanti P, Pennacchio F (2002) Metabolic and symbiotic interactions in amino acid pools of the pea aphid, *Acyrthosiphon pisum*, parasitized by the braconid *Aphidius ervi*. J Insect Physiol 48:507–516

Rohne O (2002) Effect of temperature and host stage on performance of *Aphelinus varipes* Forster (Hym., Aphelinidae) parasitizing the cotton aphid, *Aphis gossypii* Glover (Hom., Aphididae). J Appl Entomol Zeitschrift für Angewandte Entomologie 126:572–576

Shigenobu S, Watanabe H, Hattori M, Sakaki Y, Ishikawa H (2000) Genome sequence of the endocellular bacterial symbiont of aphids *Buchnera* sp. APS. Nature 407:81–86

Sigsgaard L (2000) The temperature-dependent duration of development and parasitism of three cereal aphid parasitoids, *Aphidius ervi*, *A. rhopalosiphi*, and *Praon volucre*. Entomol Exp Appl 95:173–184

Simon JC, Rispe C, Sunnucks P (2002) Ecology and evolution of sex in aphids. Trends Ecol Evol 17:34–39

Stacey DA, Fellowes ME (2002) Influence of elevated CO_2 on interspecfic interactions at higher trophic levels. Glob Chang Biol 8:668–678

Sullivan DJ, Völkl W (1999) Hyperparasitism: multitrophic ecology and behaviour. Ann Rev Entomol 44:291–315

Terblanche JS, Deere JA, Clusella-Trullas S, Janion C, Chown SL (2007) Critical thermal limits depend on methodological context. Proc R Soc B Biol Sci 274:2935–2942

Thomas MB, Blanford S (2003) Thermal biology in insect-parasite interactions. Trends Ecol Evol 18:344–350

Traugott M, Bell JR, Broad GR, Powell FJ, van Veen FJF, Vollhardt JMG, Symondson WOC (2008) Endoparasitism in cereal aphids: molecular analysis of a whole parasitoid community. Mol Ecol 17:3928–3938

Tsuchida T, Koga R, Fukatsu T (2004) Host-plant specialization governed by facultative symbiont. Science 303:1989

van der Putten WH, de Ruiterb PC, Bezemera TM, Harvey JA, Wassen M, Wolters V (2004) Trophic interactions in a changing world. Basic Appl Ecol 5:487–494

Van Veen FJF, Mueller CB, Pell JK (2008) Food web structure of three guilds of natural enemies: predators, parasitoids and pathogens of aphids. J Anim Ecol 77:191–200

Vansteenis MJ, Elkhawass KAMH (1995) Life-history of *Aphis-gossypii* on cucumber – influence of temperature, host-plant and parasitism. Entomol Exp Appl 76:121–131

Walton MP, Powell W, Loxdale HD, Allen-Williams L (1990) Electrophoresis as a tool for estimating levels of hymenopterous parasitism in field populations of the cereal aphid. Entomol Exp Appl 54:271–279

Yu DS, van Achterberg K, Horstmann K (2005) World Ichneumonoidea 2004. Taxonomy, biology, morphology and distribution. CD/DVD. Taxapad, Vancouver, Canada. http://www.taxapad.com

Zamani AA, Talebi A, Fathipour Y (2007) Effect of temperature on life history of *Aphidius colemani* and *Aphidius matricariae* (Hymenoptera: Braconidae), two parasitoids of *Aphis gossypii* and *Myzus persicae* (Homoptera: Aphididae). Environ Entomol 36:263–271

Chapter 5
Implications of Climate Change for Cereal Aphids on the Great Plains of North America

J.P. Michaud

Abstract The Great Plains of North America represent a vast expanse of fertile land characterized by extreme weather conditions, generally limited rainfall, and long distance migration of insect pests, including aphid species that are especially damaging to graminaceous crops. An increase in annual mean temperatures would have certain predictable consequences for cereal aphid populations via direct effects on aphid biology, and indirectly through effects on plants and natural enemies. However, any sustained shift in prevailing wind patterns associated with atmospheric warming could generate an unpredictable cascade of ecological consequences for both agriculture and cereal aphids, mediated largely by changes in rainfall patterns and migration pathways, respectively. The direct effects of warming would be generally beneficial for aphids through most of the year, except in summer when high temperatures are already limiting to survival in much of the region. Additional indirect effects will be mediated by changes in the phenology of cereal crops and wild host plants, adjustment of agronomic practices, and likely impacts on natural enemies. Possible adverse consequences for agriculture include northern range expansions of pest aphid species, prolonged periods of asexual reproduction in autumn, improved survival of overwintering eggs, earlier development of spring populations, and more efficient virus transmission over greater distances.

Keywords *Diuraphis noxia* · *Rhopalosiphum padi* · *Schizaphis graminum* · Tillage · Drought · *Sorghum bicolor* · *Triticum aestivum*

5.1 Introduction

Temperature has long been recognized as a critical factor influencing aphid life history and polyphenism. Many ecological studies have demonstrated correlations between aphid population dynamics and seasonal temperature cycles. While

J.P. Michaud (✉)
Department of Entomology, Kansas State University, Manhattan, KS, USA
e-mail: jpmi@ksu.edu

P. Kindlmann et al. (eds.), *Aphid Biodiversity under Environmental Change*,
DOI 10.1007/978-90-481-8601-3_5, © Springer Science+Business Media B.V. 2010

warming trends can accelerate aphid population growth, extremes of temperature, both hot and cold, can be important limiting factors, and also affect the survival and effectiveness of aphid natural enemies. Although resistance to aphids has been selectively bred into certain crops, there are cases in which its expression is temperature-dependent. Furthermore, the pest status of some aphids derives specifically from their role in vectoring plant viruses and the acquisition and transmission of these can vary with temperature. Given general acceptance of a continuing trend toward global atmospheric warming, we might wisely consider what impact this might have on aphid species with importance as crop pests. In reference to cereal aphid management on the Great Plains, Peairs et al. (2005) stated the following: "Climate and weather often exhibit greater influence on pest activity than can be achieved with available pest management practices." Within the context of climate change, positive, negative, and null responses have all been inferred for aphids (Newman 2006). In this chapter, I address the potential implications of climate change for cereal aphids on the Great Plains of North America. This region represents the 'bread basket' of cereal production in North America and is thus of great agricultural importance. Moreover, most of the aphids damaging cereal crops have been extensively studied and ample data is available on their biology and ecology that can be interpreted within a context of climate change.

Many localized scenarios of climate change can be generated from assumptions of global warming and it is not feasible to explore them all in the context of aphid ecology. Rather, I will focus on a particular projection of large-scale changes that could be of specific relevance to Great Plains agriculture and its aphid pest fauna on an area-wide basis. Fu et al. (2006) used 27 years of satellite data to demonstrate global expansion of the semi-arid tropical circulation by about 2° latitude over this period. A function of tropospheric warming between latitudes of 15° and 45°, this phenomenon was equally pronounced during warm and cool seasons. It is characterized by rising air pressures at sea level around 30° latitude and a shifting of the jet streams polewards in both northern and southern hemispheres. Since the Great Plains occupy an inland region spanning 30°–50°N latitude, they lie directly in the path of semi-arid tropical expansion in North America, a scenario that would likely result in increased seasonal mean temperatures and reduced average precipitation. Therefore, I examined the potential implications of a scenario of warmer, shorter winters, hotter, longer summers, and generally drier conditions, for cereal aphid ecology in the Great Plains region.

5.2 The Region and Its Climate

The Great Plains of North America is a vast, central plateau that extends across Texas, Oklahoma and portions of New Mexico in the south, through eastern Colorado, Wyoming, Kansas and Nebraska, and northward through the Dakotas and Montana into the Canadian prairie provinces of Manitoba, Saskatchewan and Alberta. The eastern border comprises portions of Minnesota, Iowa, Illinois and

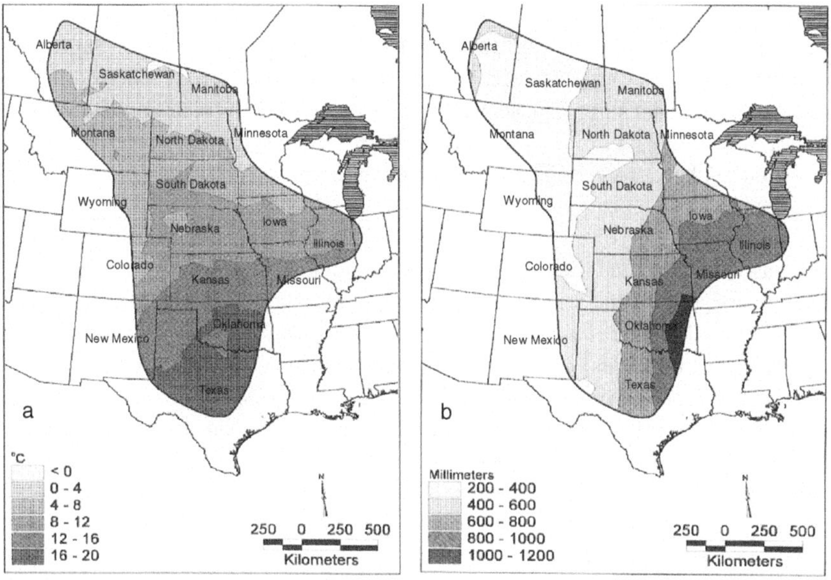

Fig. 5.1 Map of the great plains region depicting spatial distributions of mean annual temperature (**a**) and rainfall (**b**) (reprinted with permission from Leemans and Cramer 1991)

Missouri. Current mean annual temperatures throughout the region are depicted in Fig. 5.1a. Arid, semi-desert conditions prevail along the southern and western edges of this plateau, gradually giving way to sweeping prairie grasslands of higher agronomic potential to the north and east. The plateau comprises the former bed of a shallow, inland sea in which a great range of soil types are represented, many rich in organic matter and high in fertility. The region is characterized by a relatively temperate climate with moderate to low precipitation, especially along the western side that experiences a 'rain shadow' effect from the Rocky Mountains (Fig. 5.1b). During spring and early summer, prevailing winds from the south bring warm, moisture-laden air from the Gulf of Mexico to mix with cooler air from the north, often generating dramatic 'super cell' thunderstorms and tornados for which the region is renowned. In a region where agronomy is limited more by water availability than by any other physical factor, these weather patterns bring much-needed rainfall during a period that is critical both for the maturation of fall-planted crops, and for the germination of those planted in summer. These strong, southerly winds can persist for days at a time and facilitate the migration of a large number of migratory insect pests, including aphids, that inflict economic losses on agricultural crops at much more northerly latitudes than permit their overwintering survival. For example, Rogers et al. (1972) reviewed long term climatology data for Oklahoma and found that winds were southerly 48.1% of the time. The same authors estimated that alatae of the greenbug, *Schizaphis graminum* (Rondani), could theoretically be transported 280 km in a normal day. A variety of multivoltine insects are thus able to amplify their populations on early planted summer crops in the south and, utilizing these winds, migrate northward in large numbers to exploit the same crops at higher latitudes where they are planted at later dates and have delayed phenology. Any significant shift in these prevailing weather patterns resulting from climate change would have a dramatic impact on both seasonal rainfall and insect migration routes, likely triggering unpredictable cascades of ecological and agronomic consequences.

Cycles of temperature span extreme values in the Great Plains on both diurnal and seasonal scales and thermal impacts on both plants and insects are often amplified by persistent strong winds. The prairie environment has a remarkable capacity for rapid and dramatic changes in weather, often triggered simply by an abrupt shift in wind direction. Thus, resident insect species endure scorching temperatures in summer and wind-driven, sub-zero cold in winter, often with little snow cover, and must also tolerate short-term temperature fluctuations that can span an amplitude of 20–25°C within a 24 h period. These rapid shifts in temperature, often combined with powerful winds, are conditions difficult for woody shrubs and trees to tolerate and are largely responsible for retention of the prairie ecosystem in early successional stages dominated by grasses and herbaceous plants. The relative absence of significant numbers of trees on the Great Plains may be one factor limiting the importance of certain obligately heteroecious cereal aphids such as *Metopolophium dirhodum* (Walker) that, although present, does not achieve the pest status it enjoys in other cereal-producing regions. Although both cereal aphids and their natural enemies have evolved a variety of physiological and behavioral mechanisms for tolerating rapid fluctuations in temperature, a shift to higher mean temperatures and/or more arid conditions would not impact all species equally and arthropod community structure could be dramatically and permanently altered, possibly with serious agronomic consequences.

5.3 Agroecosystems

The agronomy of the Great Plains is largely based on cattle ranching in combination with various rotations of cereal, forage and oilseed crops, grown for the most part in large monocultures, an agricultural configuration that generally favors insect pests such as aphids that have high rates of colonization (Bottenberg and Irwin 1992; Ogenga-Latigo et al. 1992; Elliott et al. 1999; Favret and Voegtlin 2001). Winter wheat, *Triticum aestivum* L., remains the predominant cool season crop and is still grown without rotation on substantial acreage as 'continuous wheat' despite frequent recommendations to the contrary by agronomic and pest management professionals. Maize remains one of the most profitable summer cereals under irrigation, and in regions where seasonal rainfall is adequate, being replaced largely by sorghum, *Sorghum bicolor* (L.), in drier southern areas, and areas where aquifer depletion has rendered irrigation unsustainable. Cotton, *Gossypium hirsutum* L., is widely grown in the south, and soybeans, sunflowers and alfalfa are alternative crops that may be grown with or without irrigation.

Tillage practices are foremost among the cultural activities likely to have significant, area-wide impacts on cereal aphid ecology. Crop residues increase the snow cover retained in fields, raising winter soil temperatures, and lowering spring soil temperatures as a function of increased soil moisture (Olfert et al. 2002). Although adoption of 'no-till' and 'minimum till' agriculture continues to increase on the Great Plains, as of this writing a significant proportion of farmers still employ conventional tillage. The primary incentive for reduced tillage is enhanced conservation of soil moisture in regions where strong solar radiation combines with hot summer winds to quickly desiccate disturbed soils. An additional benefit is reduced soil erosion by wind and water runoff. Interestingly, Bernacchi et al. (2006) estimated that complete conversion of the corn/soybean agroecosystem in the USA to no-till agriculture would offset about 2% of annual carbon emissions via improved carbon sequestration in the soil. Generally, any changes in the cropping system that improve its resemblance to a more natural habitat would be expected to reduce rates of aphid colonization (Favret and Voegtlin 2001). Although hard evidence is rather sparse, there is a general consensus that no-till farming generally reduces aphid problems (Burton et al. 1987; Borowy 2004), although not necessarily in all situations (Rice and Wilde 1991; Hesler and Berg 2003; Sempruch et al. 2007). Lower aphid populations may result from reduced colonization of the crop, likely because crop residues function as a sort of mulch that reduces the contrast of young plants against bare soil (Smith 1969; 1976; Burton and Krenzer 1985; Heimbach et al. 2004). Furthermore, natural enemy activity in fields may be enhanced due to the increased shelter afforded by crop residues and the alternative food sources they may indirectly provide (House and Parmelee 1985; French et al. 1998). There is also evidence that the community of decomposers in the soil, known to be beneficiaries of reduced tillage practices, can influence cereal aphid abundance via their effects on mobilization of nitrogen from soil organic matter (Ke and Scheu 2008).

Notwithstanding the numerous agronomic and environmental benefits of no-till agriculture, many farmers have been resistant to change their tillage practices. Although the cost of equipment conversion has posed an entry-level barrier to some, increasing fuel costs have recently generated an additional economic incentive for reducing the frequency of vehicular traffic on arable land and can be expected to spur more wide-spread adoption of no-till farming in the future. Herbicides are increasingly substituted for mechanical weed control in so-called 'chemfallow' programs and can be applied in wide swaths with light equipment, generating substantial fuel savings. In addition, some modern equipment combinations permit planting, fertilization, and pre-emergent herbicide applications all in a single pass over undisturbed soil, further reducing fuel consumption. Thus, any changes in the nature and intensity of cereal aphid infestations driven by changing climate will occur against a backdrop of shifting agricultural practices driven, in turn, by rising input costs and the increasing importance of improved water and soil conservation to both stakeholders and the general public.

5.4 Aphid Populations

Although aphids can disperse passively on the wind over vast distances, the number of migrants and their survival rates are strongly affected by climate. For example, Sparks et al. (2007) linked increases in the numbers of species of migratory Lepidoptera arriving in southern UK to warming temperatures in southwestern Europe. Thus, one concern connected to climate change is the potential for range expansion of tropical aphid species into more temperate latitudes. For example, the yellow sugarcane aphid, *Sipha flava* (Forbes), is an indigenous species that poses a potential risk to Great Plains cereal production in this regard. It injects a feeding toxin, can be devastating to a wide range of cereal crops, and has few effective natural enemies (Starks and Mayo 1985; Hentz and Nuessly 2004). However, it is a subtropical species that is normally an economic concern only in the Gulf Coast states. It has only rarely produced outbreak populations as far north as Kansas, but it is a prime candidate for range expansion northward into the Great Plains under a climate warming scenario. Although capable of producing sexual morphs and overwintering eggs, its status as a serious pest may well be linked to year-round asexual reproduction that presently occurs only in southern latitudes.

Rather than attempting to catalogue exotic aphid species that might invade the Great Plains under scenarios of a warming climate, the remainder of this chapter will focus on the potential implications for three aphid species currently recognized as important pests of cereal crops in the region, the bird cherry-oat aphid, *Rhopalosiphum padi* L., the Russian wheat aphid, *Diuraphis noxia* Mordvilko, and the greenbug, *S. graminum*. Both *D. noxia* and *S. graminum* cause chlorotic lesions on wheat that can potentially kill plants or severely reduce grain yield, and the latter species can also damage sorghum. Feeding damage by *R. padi* on wheat is less conspicuous, but can be significant (Roza-Gomes et al. 2008) and this species is also the primary vector of barley yellow dwarf virus (BYDV), a serious disease of wheat and barley. All three species can utilize a wide range of wild grass hosts in the family Poaceae, but normally develop large populations only on cultivated grain crops. Both Russian wheat aphid and bird cherry-oat aphid can be considered cool season aphids that perform best on cool season grasses, whereas the greenbug is more of a warm season aphid able to thrive at higher temperatures. Michels and Behle (1989) examined the life histories of all three species under three temperature regimes of similar mean: 10°C (constant), 5–15°C (night-day) and 1–19°C (night-day). Aphid responses to these temperatures were very species-specific and revealed that the birth rates of *S. graminum* and *R. padi* both benefited from a wider fluctuation of temperature than did *D. noxia*. Similarly, Kieckhefer and Elliott (1989) observed that lower temperature regimes favored the development and reproduction of *D. noxia* in comparison to the higher temperature regimes they tested. Harvey and Martin (1988) used field cages to demonstrate that *D. noxia* is better able to survive winter conditions in Kansas than is *S. graminum*. Thus, despite sharing many common host plants, these species are ecologically distinct and are likely to respond to climate change in ways that are in some cases similar, but in others, quite species-specific.

5.4.1 Bird Cherry-Oat Aphid

Among cereal aphids on the Great Plains, *R. padi* is generally the first to become abundant early in spring, and the last to be active late in autumn, mirroring its phenology in western France (Dedryver 1978). If overwintering egg mortality occurs at a relatively constant rate throughout winter, as suggested by Leather (1980), a shorter winter would likely improve the egg survival of holocyclic clones in more northern latitudes. Certain models have correlated warm periods of winter with increases in spring populations of *R. padi* in cereals in France (Pierre 1987). However, Griffiths and Wratten (1979) demonstrated considerable variation in low-temperature tolerance among various clones of *R. padi* that correlated with winter temperatures at their sites of origin, indicative of populations adapted to local climatic conditions. Ma and Ma (2007) examined *R. padi* responses to thermal gradients and demonstrated its capacity to acclimate to high temperatures. It is also notable that *R. padi* has a behavioral mechanism for countering extremes of temperature – it can descend to below the soil surface and feed at the very base of grass stalks (Wiktelius 1987). Kieckhefer et al. (1974) reported that it was unable to overwinter asexually in the Great Plains much further north than Nebraska, but it would be worthwhile to retest present day populations.

Aphids such as *R. padi* vary greatly in their dependence on sexual reproduction in different geographic regions and comparisons of different clones typically reveal tremendous intraspecific variation in expression of sexuality, both in Europe (Hulle et al. 1999) and in North America (Voegtlin and Halbert 1998). Bale et al. (1999) have demonstrated the existence of *R. padi* clones that employ mixed strategies, producing sexual forms while continuing asexual reproduction and Rispe et al. (1999) have demonstrated similar variable life cycles among European *R. padi* clones. The ability of predominantly asexual clones to resort to sexuality as needed, and periodically receive gene flow from holocyclic populations (Halkett et al. 2008), are traits that should predispose this species to high rates of adaptation in the face of environmental perturbations.

One of the first authors to consider the implications of global warming for aphid life histories reported on improved survival of *R. padi* overwintering in Poland, earlier hatching of overwintered eggs, and faster development of immatures in spring (Ruszkowska 1999, 2002, 2005). Thus, another effect could be earlier initiation of population increase in spring due to faster accumulation of degree-days and earlier transition across the low temperature thresholds for development and reproduction. Increased seasonal mean temperatures would likely enable year-round anholocyclic reproduction to occur at higher latitudes than is now possible. Continuous asexual reproduction of *R. padi* through mild winters was known to occur in Hungary (Kuroli 1984) and has recently been reported from western Poland (Ruszkowska 2006) where colonies now remain active on winter cereals into January (Gaezewski 2007). Even in holocyclic forms, periods of parthenogenetic reproduction could be extended during warm autumns, increasing the period of aphid activity on winter cereals in their seedling stages. Although *R. padi* is a host-alternating aphid that overwinters on wild *Prunus* sp. in the northern parts of its range, it enjoys year-round, anholocyclic reproduction on cereals throughout southern and central latitudes of the Great Plains and could likely employ this strategy at ever higher latitudes under a warming scenario, leading to increased transmission of BYDV and other cereal viruses in these regions.

5.4.2 Russian Wheat Aphid

Even though the Russian wheat aphid is adapted to cool season conditions, low temperatures can be limiting to its population growth in fall and spring throughout much of its range. Although *D. noxia* oviparae and eggs have been recovered in North America occasionally (Kiriac et al. 1990; Stoetzel and Hammon 1992) complete holocyclic reproduction is not thought to occur with any regularity in North American clones, even though it is reportedly common in clones from Europe and Asia (Starý 1996; Zhang et al. 2001). The lack of a holocycle may currently aid in limiting its northern range in North America and could, in theory, retard the rate at which new biotypes are able to evolve.

Hammon and Peairs (1992) repeatedly found *D. noxia* selectively colonizing wheat on the southern aspects of east–west furrows and irrigation ditches in Colorado, locations where temperatures were consistently higher early in spring and late in fall. The pest status of *D. noxia* on the Canadian prairies has so far been negligible, largely due to poor winter survival. Butts (1992) observed general population decline with surface temperatures between 0 and –10°C, although low levels of survival were noted following the winter of 1989–1990 when temperatures failed to drop below –25°C. Similarly, Butts and Schaalje (1997) concluded that *D. noxia* can overwinter successfully in the northern US and Canada when snow cover maintains temperatures between zero and –5°C, but exposure to temperatures below –10°C lead to dramatic population decline. Winter temperatures are also limiting to *D. noxia* survival at higher elevations in more southern latitudes (Messina 1993). If mean winter temperatures were to rise, range expansion by *D. noxia* would likely occur northward in the higher latitudes of the Great Plains and westward into the higher elevation plateaus in Utah, Nevada and Washington states. Whether these range expansions would have economic impact on wheat and barley production would likely hinge on the ability of local natural enemy populations to respond to the availability of *D. noxia* as a novel prey/host, but it is noteworthy that wheat and barley cultivars with *D. noxia* resistance have not been developed for these localities. Natural range expansion of *D. noxia* is thought to have occurred from Turkey through the Mediterranean region and northward through central Europe during the late 1980s (Lukášová et al. 1999). Its presence in Kenya (Malinga et al. 2007) indicates its potential for adaptation to tropical conditions.

Although the presence of snow cover tends to buffer the impact of low temperatures on both aphids and wheat, fall infestation by *D. noxia* can reduce the cold hardiness of wheat and thus exacerbate problems with winterkill in northern regions (Thomas and Butts 1990). Were climate changes to result in winter temperature fluctuations of higher amplitude in the northern Great Plains, periodic mid-winter snow melts could expose more aphids to freezing mortality when temperatures subsequently drop, and more of the wheat to winterkill. However, surviving aphids could exploit even short periods of warmth to increase in numbers, whereas wheat would not accomplish significant compensatory growth until emergence from dormancy.

5.4.3 Greenbug

The greenbug is essentially a warm season aphid with maximal rates of population increase typically reported around 30°C (Kirkland et al. 1981). It is able to reproduce asexually year-round throughout much of the southern portion of the Great Plains (Shufran et al. 1997), but winter temperatures can be limiting to its survival as far south as Oklahoma (Arnold 1981). Many 'biotypes' of greenbug have been distinguished on the basis of differential plant responses (Ullah and Peters 1996), and various strains can be differentiated based on photoperiodic responses (Mittler and Gorder 1991), and insecticide resistance (Rider and Wilde 1998). The periodic occurrence of a sexual cycle is thought to facilitate evolution of the extensive biotypic variation observed in this aphid (Shufran and Wilde 1994; Puterka and Peters 1995). The evolutionary plasticity of greenbug is further evidenced by its history of host range expansion. Although first described in Italy, its region of origin is uncertain and it may well have been present in North America prior to wide-spread wheat cultivation. It emerged rather suddenly as a pest of wheat in the Great Plains in the late 1800s (Hunter 1909), and abruptly began to colonize sorghum in the late 1960s (Harvey and Hackerott 1969). The ability to utilize sorghum created a substantial 'food bridge' for greenbugs during the summer period when wheat is not available and populations are otherwise restricted to wild grass hosts that rarely support large populations (Anstead et al. 2003).

Currently, economic greenbug problems in Great Plains wheat are largely confined to the panhandle regions of Oklahoma and Texas. When the early wheat cultivars in this region mature, massive swarms of alate migrants can move northward to attack wheat in earlier phenological stages (Daniels 1977a, b), occasionally as far north as Canada (Irwin and Thresh 1988). However, overwintering is a challenge for the greenbug in Kansas (Harvey and Martin 1988), and appears unlikely to occur north of Nebraska (Kieckhefer et al. 1974). In Kansas, the relative contribution of migrant alates vs. overwintered eggs to local greenbug problems wheat has long been the subject of debate, but in the absence of significant migrant swarms over the past 10 years, greenbug has rarely caused significant losses in Kansas wheat, one exception being areas along the southern border with Oklahoma during the spring of 2006, following an exceptionally warm winter. It is therefore tempting to conclude that warmer winters would enable greenbug populations to become more damaging to winter wheat at higher latitudes that, as yet, do not support year-round populations.

5.4.4 Shorter, Warmer Winters

Menzel et al. (2006) analyzed the historical phenology of 542 plant and 19 animal species in 21 European countries and concluded that the seasonal onset of spring/summer had advanced by 6–8 days over the past 30 years, and that autumn was beginning about 3 days later. Newman (2006) modeled climate change consequences for cereal aphids in Canada and concluded that, although there seemed to be no common response among aphid species, temperature was the single variable responsible for driving the variety of aphid responses observed. Any significant rise in seasonal minimum temperatures during winter would have several predictable effects on cereal aphid populations. The first would be a reduction in winter mortality wherever aphid survival is limited by extremes of cold temperature and this could facilitate northern range expansion by a number of species, perhaps with most impact on the Canadian prairies that normally experience the most severe winter conditions. Many effects of warmer winters on cereal aphid populations might be also mediated indirectly by climate effects on winter crop phenology and, ultimately, by climate-driven changes in sowing dates and cultivar selections. Dixon (2000) reviewed suction trap data on the English grain aphid, *Sitobion avenae* (F.) and observed that larger spring flights correlated with preceding winters that were mild, but pointed out that such flights were not solely responsible for injurious populations developing in the wheat that year. Rather, summer aphid populations were favored by conditions that delayed wheat maturity and provided a longer time window for aphid population development. However, a trend toward shorter winters would enable farmers to shift to shorter season wheat cultivars at a given latitude, negating at least partially any increases in the length of period that winter wheat remains vulnerable to aphid attack.

5.4.5 Longer, Hotter Summers

The consequences for cereal aphids of increased summer heat intensity and length of season are more difficult to anticipate than those associated with milder winters. Aphid reproduction and population growth rates generally demonstrate parabolic relationships with temperature, increasing at a diminishing rate to some maximum and then abruptly decreasing (Dixon 2000). Thus, in regions experiencing marked extremes of temperature such as the Great Plains, the prime seasons for aphid population growth are spring and fall when generally moderate thermal conditions prevail. High temperatures are the primary physical factor limiting aphid populations in summer throughout much of the Great Plains and the projected trend of climate change would likely generate an even longer aphid-free period for summer crops in southern latitudes. Newman (2006) used a series of climate models to project changes in abundance and geographic range of cereal aphid populations in Canada and inferred both longitudinal and latitudinal shifts would result, increasing summer populations in coastal regions and reducing them in continental regions such as the northern Great Plains. Another possibility might simply be earlier onset of aphid infestations in spring, and later onset in fall, with little effect on overall population levels, provided cycles of natural enemy populations followed similar patterns. However, summer also poses a survival challenge for many aphid natural enemies as it represents a period of harsh conditions and food scarcity without the option of hibernation and an extension of this period could be detrimental to these species.

Most studies suggest that the optimal temperature for *R. padi* population growth is in the neighborhood of 20°C, but clones exist that are adapted to much higher temperature conditions. For example, Asin and Pons (2001) described a high-temperature adapted population from the arid northeastern Iberian peninsula with a maximum population growth rate at 28.5°C. Similarly, de Barro and Maelzer (1993) found that air temperatures of 36°C were required to directly reduce survival of a population of *R. padi* colonizing perennial grasses in South Australia. Thus, particular lineages of *R. padi* have demonstrated the ability to adapt to high temperature conditions in some very arid regions. Earlier *R. padi* activity in spring would increase the risk of early virus transmission to spring sown cereals at higher latitudes. Low temperature is one of the more important cues used by aphids to terminate parthenogenetic reproduction and initiate production of sexual forms (Ward et al. 1984). Thus, a lengthening of the fall season could eventually increase the number of asexual generations of *R. padi*, leading to extended periods of virus transmission to winter cereals in early growth stages.

Hot summer weather can be critically limiting to greenbug populations. This was illustrated by the long term historical analysis of greenbug population dynamics in relation to climatological conditions in Oklahoma carried out by Rogers et al. (1972). The authors concluded that outbreak populations were most often preceded by above-normal temperatures in winter, spring and autumn, and below normal temperatures in summer. High temperatures tend to increase the proportion of greenbug nymphs developing into alatae (Mayo and Starks 1974), an effect likely to reduce the persistence and size of greenbug colonies. Temperature can affect greenbug preference for particular sorghum cultivars (Starks et al. 1973) and particular sources of sorghum resistance to greenbug have been shown to be more effective at higher, rather than lower, temperatures (Schweissing and Wilde 1979; Harvey et al. 1994; Thindwa and Teetes 1991). Similarly, Salas and Corcuera (1991) concluded that barley seedlings grown at high temperatures and long photoperiods were most resistant to greenbug because these conditions favored elevated gramine concentrations in the youngest leaves. Nonetheless, greenbugs can take advantage of microclimatic effects within the folded and overlapping leaves of sorghum plants under high temperature conditions. Susidko and Skylar (1974) found that greenbugs within sorghum foliage could benefit from conditions of 20–22°C and 65–75% RH even when ambient conditions were 35°C and 34% RH. Even though sorghum seedlings are highly susceptible to greenbug feeding, colony establishment is difficult on young sorghum plants because these do not afford adequate protection from punishing winds and rain that can be important aphid mortality factors.

The survival of *D. noxia* in the heat of summer already tends be poor in southern states such as Texas (Archer and Bynum 1993). Increases in either the period or intensity of summer heat in the southern Great Plains could actually result in the disappearance of *D. noxia* from some southern portions of its range. Possibly countering this outcome is an expansion of the distribution of the RWA2 biotype that was first detected in Colorado in spring of 2003 (Haley et al. 2004). In addition to overcoming resistance based on the *Dn*4 gene, the primary source of resistance in North American commercial wheat cultivars, RWA2 thrives under higher temperature conditions than the original North American biotype, and causes plant damage symptoms to develop at faster rates, especially at higher temperatures (Jyoti et al. 2006; Michaud et al. 2006). More than a decade earlier, Webster et al. (1993) had noted higher fecundity in populations of *D. noxia* collected in South Africa compared to those from Oklahoma when reared at 25°C, a difference not evident at 15°C. Thus there are indications that *D. noxia* also has an innate capacity to adapt to warmer climatic conditions.

5.4.6 Drought

Summer on the Great Plains is often a period of prolonged drought, a situation likely exacerbated under the proposed scenario of climate change. Under arid conditions, the synergistic interactions between drought stress and aphid feeding pose additional concerns to cereal production. Aphids remove large amounts of water and dissolved photosynthate from the plant phloem, increasing transpiration rates and leaf surface temperatures. These effects can be especially damaging to plant productivity when soil moisture is limiting. Infestation of wheat by *D. noxia* leads to chlorosis followed by plant desiccation, which in turn triggers the induction of wing development in nymphs (Baugh and Phillips 1991). Drought may increase plant sensitivity to aphid feeding or decrease the effectiveness of specific resistance traits. Especially damaging infestations of *D. noxia* in barley and wheat have been associated with hot, dry weather, particularly during the ripening stage (Miller et al. 1992; Starý and Lukášová 2002). Similarly, high populations of greenbug on cereals in Texas and Oklahoma have been correlated with below-normal rainfall during winter months (Wallin and Loonan 1977) and irrigation has shown a negative correlation with aphid populations on barley (Hamam and Salman 2007).

Drought stressed plants may be less resistant to aphid feeding, and less able to compensate for its effects. For example, Dorschner et al. (1986) showed that infestation by *D. noxia* decreased the ability of wheat to tolerate drought stress and Riedell (1989) demonstrated much the same effect on barley. Many studies have shown that aphid population growth can be favored on drought-stressed plants (Dorschner et al. 1986; Liang et al. 1998; Kelm and Klukowski 2000; McVean and Dixon 2001), although others have measured negative effects on aphid reproduction (Sumner et al. 1986; Behle and Michels 1988; Pons and Tatchell 1995). Similarly, greenbug infestation can add to drought stress in wheat, raising plant temperatures above levels suitable for growth. Cabrera et al. (1995) showed than greenbug feeding caused phytochemical changes in barley similar to those induced by drought stress. The use of infrared transducers to sense greenbug infestations remotely in winter wheat (Michels et al. 1999) have been complicated by the fact both greenbug feeding and drought stress produce relatively warmer plants in comparison with ambient temperature. However, Li et al. (2008) showed that greenbug feeding on sorghum exacerbated water stress and associated yield losses. Thus, any trend toward increasing aridity would likely serve to exacerbate the negative impact of cereal aphids on crop plants and impede plant ability to recover from aphid feeding damage.

5.4.7 Virus Transmission

The importance of long range migration in the annual re-infestation of northern regions of the Great Plains by cereal aphids that persist largely in parthenogenetic populations to the south has been emphasized by many authors (Kieckhefer et al. 1974; Irwin and Thresh 1988), but less attention has been paid to the attendant movement of aphid-borne viruses. For example, long-distance migration of alate greenbug in China has been implicated in BYDV transmission to spring wheat (Dong et al. 1995). Should climate warming permit migrant alatae to begin colonizing more northern regions of the Great Plains, the risk of long-distance transmission of aphid-borne viruses to these areas would increase.

BYDV is one of the most serious diseases of wheat, oats and barley and epidemics are largely dependent on outbreak populations of its aphid vectors, primarily *R. padi* (Halbert and Pike 1985; Chapin et al. 2001). Apterae tend to be more efficient than alatae in transmitting BYDV (Guo and Moreau 1996), and the establishment of successful colonies by migrant alatae is necessary for local infections to spread. Ruszkowska (2005) has documented increased transmission of BYDV by *R. padi* in cereal crops in Poland since the onset of shorter, milder winters associated with climate change in central Europe. The aphid is now able to overwinter on winter cereals in parthenogenetic form and the frequency of BYDV infections has consequently increased. There is also a tendency for interplant movement of *R. padi* apterae to increase at higher temperatures and result in higher rates of BYDV infection among adjacent wheat plants (Smyrnioudis et al. 2000). Sadeghi and Dedryver (2004) compared 16 clones of *R. padi* for their transmission of two BYDV isolates and found that transmission percentage increased with temperature from 15 to 20°C for all clones. Similarly, Lucio-Zavaleta et al. (2001) found increasing efficiency of BYDV transmission when acquisition and inoculation occurred at higher temperatures. Lowles et al. (1996) associated increased BYDV infection rates with increases in autumn temperatures in the UK. Blackmer and Bishop (1991) studied *R. padi* and BYDV infection of maize in southwestern Idaho for a period of 5 years and observed that temperature cycles during winter months differed between epidemic and non-epidemic years, the former correlating with warmer winter conditions. More recently, Habekuss et al. (2009) found that the lower temperature threshold for BYDV transmission in Germany was 10°C and linked longer warm periods in autumn and winter to increased rates of infection in winter barley. Therefore, there is good reason to expect climatic warming to have a net positive effect on the frequency and intensity of BYDV infections in cereal crops, both via effects on aphid behavior and polyphenism, and via direct effects on transmission efficiency.

5.4.8 Biological Control

Biological control of cereal aphids on the Great Plains has been contrasted to that occurring in Europe and less arid regions. For example, a relatively low abundance of spiders is evident in the arid western regions (Brewer and Elliott 2004). Aphid natural enemies may be negatively affected by extremes of high temperature, or their impact on aphid populations may be reduced. Although the developmental time of coccinellids is shortened under warmer thermal regimes, consumption of aphids may actually be decreased because the feeding period is reduced (Michels and Bateman 1986). High temperatures accelerate rates of water loss through the insect cuticle and such losses are more pronounced during periods of active foraging. Aphids tend to be very patchily distributed and aphidophagous insects must expend considerable time and effort in actively searching for them. Hot weather renders predators and parasitoids susceptible to desiccation, forcing them to curtail foraging activities during much of the day. Many insects rely on drinking dew from plants during early morning hours, but this is not a reliable source of moisture on the Great Plains. Nectar feeding is well-documented among parasitoids, but flowers are not always available and local competition for nectar sources can be high. Consequently, many insects on the High Plains, including aphidophagous coccinellids, lacewings and predatory wasps, obtain hydration from particular plants such as sunflower, *Helianthus annuus* L., that bear extrafloral nectaries (Michaud and Qureshi 2005).

Although many beneficial insects adapted to the Great Plains environment may effectively rely on a variety of plants as sources of water and supplementary foods, it is often low prey availability that limits their ability to survive summer months. Throughout southern and central latitudes of the Great Plains, the heat of summer normally maintains aphid populations at very low numbers in all but isolated situations and the result is a paucity of food for the natural enemies that rely on them. Parasitoids of the families Aphidiidae and Aphelinidae are obligate parasitoids of aphids that can only survive as adults for limited periods without access to host aphids. Similarly, most aphidophagous syrphids and lady beetles rely on certain aphid species as food sources essential for completing their life cycles. Although polyphagous coccinellids may resort to alternative, non-aphid prey for sustenance, these often do not support reproduction, though they may extend lifespan (Hodek and Honěk 1996). In addition, when developing on an exclusive diet of pollen, *Coleomegilla maculata* (DeGeer) larvae have a greatly increased water demand (Michaud and Grant 2005). In response to food limitation, *Hippodamia convergens* Guérin-Méneville enters a reproductive diapause that can be broken within 3–4 days if an adequate supply of suitable aphids is encountered (Michaud and Qureshi 2005). The benefit of delayed onset of reproduction is substantially extended longevity for adult females, even though both fertility and fecundity decline gradually with age independent of their reproductive activity (Michaud and Qureshi 2006). The latter study clearly demonstrated that the costs associated with reproductive diapause increase with time, and that strength of the diapause spontaneously decays over time, suggesting that longer summers on the Great Plains could exact a higher toll on populations of this key aphid predator.

Generally, the contribution of entomopathogenic fungi to cereal aphid control on the Great Plains has been minimal due to the prevalence of arid conditions, although epizootics can develop in cereals under irrigation. For example, Feng et al. (1991) found *Pandora* (*Erynia*) *neoaphidis*, *Conidiobolous obscurus*, and *Entomophthora planchoniana* to be the most abundant fungi infesting *M. dirhodum* and *Rhopalosiphum maidis* (Fitch) in irrigated grains in Montana, but did not find these pathogens in un-irrigated fields. Wraight et al. (1993) found that infections of *D. noxia* in irrigated wheat and barley in Colorado by *P. neoaphidis* and *C. obscurus* reached 44 and 20% respectively, while infection rates in un-irrigated fields did not exceed 2.5% of aphids. Since environmental conditions for aphidophagous fungi are already poor in the region, increasing regional aridity combined with a decline in the acreage of cereals under irrigation would further diminish the significance of fungi as sources of aphid mortality. Thus, in light of our current knowledge of temperature effects on the primary aphid natural enemy groups, climate warming would be more likely to reduce than improve the overall efficacy of biological control in maintaining aphid populations below economic levels in cereal crops on the Great Plains.

Acknowledgements The author wishes to thank Drs. P. Sloderbeck and J. Reese for reading and commenting on the manuscript.

References

Anstead JA, Burd JD, Shufran KA (2003) Over-summering and biotypic diversity of *Schizaphis graminum* (Homoptera: Aphididae) populations on noncultivated grass hosts. Environ Entomol 32:662–667

Archer TL, Bynum ED Jr. (1993) Ecology of the Russian wheat aphid (Homoptera: Aphididae) on dryland winter wheat in the southern United States. J Kans Entomol Soc 66:60–68

Arnold DC (1981) Effects of cold temperatures and grazing on greenbug populations in wheat in Noble County, Oklahoma, 1975–1976. J Kans Entomol Soc 54:571–577

Asin L, Pons X (2001) Effect of high temperature on the growth and reproduction of corn aphids (Homoptera: Aphididae) and implications for their population dynamics on the northeastern Iberian peninsula. Environ Entomol 30:1127–1134

Bale JS, Block W, Somme L (1999) Clonal variability in sequences of morph production during the transition from parthenogenetic to sexual reproduction in the aphid *Rhopalosiphum padi* (Sternorrhyncha: Aphididae). Eur J Entomol 96:125–134

Baugh BA, Phillips SA Jr. (1991) Influence of population density and plant water potential on Russian wheat aphid (Homoptera: Aphididae) alate production. Environ Entomol 20:1344–1348

Behle RW, Michels GJ Jr. (1988) Responses of greenbug to drought-stressed small grain hosts. Southwest Entomol 13:55–62

Bernacchi CJ, Hollinger SE, Meyers T (2006) The conversion of the corn/soybean ecosystem to no-till agriculture may result in a carbon sink. Glob Chang Biol 11:1867–1872

Blackmer JL, Bishop GW (1991) Population dynamics of *Rhopalosiphum padi* (Homoptera: Aphididae) in corn in relation to barley yellow dwarf epidemiology in southwestern Idaho. Environ Entomol 20:166–173

Borowy A (2004) Effect of no-tillage and rye mulch on occurrence of weeds and aphids and on yields of cabbage, carrot and red beet. Acta Horticultura (ISHS) 638:147–150

Bottenberg H, Irwin ME (1992) Using mixed cropping to limit seed mottling induced by soybean mosaic virus. Plant Dis 76:304–306

Brewer MJ, Elliott NC (2004) Biological control of cereal aphids in North America and mediating effects of host plant and habitat manipulations. Ann Rev Entomol 49:219–242

Burton RL, Jones OR, Burd JD, Wicks GA, Krenzer EG Jr. (1987) Damage by greenbug (Homoptera: Aphididae) to grain sorghum as affected by tillage, surface residues, and canopy. J Econ Entomol 80:792–798

Burton RL, Krenzer EG Jr. (1985) Reduction of greenbug (*Schizaphis graminum*) (Homoptera: Aphididae) populations by surface residues in wheat tillage studies. J Econ Entomol 78:90–394

Butts RA (1992) Cold hardiness and its relationship to overwintering of the Russian wheat aphid (Homoptera: Aphididae) in southern Alberta. J Econ Entomol 84:1140–1145

Butts RA, Schaalje GB (1997) Impact of subzero temperatures on survival, longevity and natality of adult Russian wheat aphid (Homoptera: Aphididae). Environ Entomol 26:661–667

Cabrera HM, Argandona VH, Zuniga GE, Corcuera LJ (1995) Effect of infestation by aphids on the water status of barley and insect development. Phytochemistry 40:1083–1088

Chapin JW, Thomas JS, Gray SM, Smith DM, Halbert SE (2001) Seasonal abundance of aphids (Homoptera: Aphididae) in wheat and their role as Barley Yellow Dwarf Virus vectors in the South Carolina coastal plain. J Econ Entomol 94:410–421

Daniels NE (1977a) Seasonal populations and migration of the greenbug in the Texas Panhandle. Southwest Entomol 2:20–26

Daniels NE (1977b) Spring migration of the greenbug in the Texas Panhandle. Southwest Entomol 24:191–196

de Barro PJ, Maelzer DA (1993) Influence of high temperatures on the survival of *Rhopalosiphum padi* (L.) (Hemiptera: Aphididae) in the irrigated perennial grass pastures in South Australia. Aust J Zool 41:123–132

Dedryver CA (1978) Biology of cereal aphids in western France. 1. Distribution and development of populations of *Sitobion avenae* F., *Metopolophium dirhodum* Wlk. and *Rhopalosiphum padi* L., from 1974 to 1977 on winter wheat in the Rennes Basin. Annales de Zoologie, Ecologie Animale 10:483–505

Dixon AFG (2000) Aphid ecology. Chapman and Hall, London

Dong QZ, Li XY, Meng QX, Zhang GX, Wei K (1995) An investigation on long distance migration of the greenbug, *Schizaphis graminum* (Rondani), in Ningxia. Acta Entomologica Sinica 38:414–420

Dorschner KW, Johnson RC, Eikenbary RD, Ryan JD (1986) Insect-plant interactions: greenbugs (Homoptera: Aphididae) disrupt acclimation of winter wheat to drought stress. Environ Entomol 15:118–121

Elliott NC, Kieckhefer RW, Lee JH, French BW (1999) Influence of within-field and landscape factors on aphid predator populations in wheat. Landscape Ecol 14:239–252

Favret C, Voegtlin DJ (2001) Migratory aphid (Hemiptera: Aphdiidae) habitat selection in agricultural and adjacent habitats. Environ Entomol 30:371–379

Feng MG, Nowierski RM, Scharen AL, Sands DC (1991) Entomopathogenic fungi (Zygomycotina: Entomophthorales) infecting cereal aphids (Homoptera: Aphididae) in Montana. Pan-Pac Entomol 7:55–64

French BW, Elliott NC, Berberet RC (1998) Reverting Conservation Reserve Program lands to wheat and livestock production: effects on ground beetle (Coleoptera: Carabidae) assemblages. Environ Entomol 27:1323–1335

Fu Q, Johanson CM, Wallace JM, Reichler T (2006) Enhanced mid-latitude tropospheric warming in satellite measurements. Science 312:1179

Gaezewski M (2007) Autumn aphids – a greater and greater threat to winter crops. Ochrona Roslin 52:50–51

Griffiths E, Wratten SD (1979) Intra- and inter-specific differences in cereal aphid low-temperature tolerance. Entomol Exp Appl 26:161–167

Guo JQ, Moreau JP (1996) Variability among aphid clones of *Rhopalosiphum padi* L. and *Sitobion avenae* Fabr. (Homoptera: Aphididae) in transmission of three PAV isolates of barley yellow dwarf viruses. Can Entomol 128:209–217

Habekuss A, Riedel C, Schliephake E, Ordon F (2009) Breeding for resistance to insect-transmitted viruses in barley – an emerging challenge due to global warming. J für Kulturpflanzen 61:53–61

Halbert SE, Pike KS (1985) Spread of Barley Yellow Dwarf Virus and relative importance of local aphid vectors in central Washington. Ann Appl Biol 107:387–395

Haley SD, Peairs FB, Walker CB, Rudolph JB, Randolph TL (2004) Occurrence of a new Russian wheat aphid biotype in Colorado. Crop Sci 44:1589–1592

Halkett F, Plantegenest M, Bonhomme J, Simon JC (2008) Gene flow between sexual and facultatively asexual lineages of an aphid species and the maintenance of reproductive mode variation. Mol Ecol 17:2998–3007

Hamam KA, Salman AMA (2007) Evaluation of spring barley genotypes to drought stress and aphid infestation. Ann Agric Sci 52:327–340

Hammon RW, Peairs FB (1992) Distribution of overwintering Russian wheat aphid (Homoptera: Aphididae) in furrow-irrigated small grains in western Colorado. J Econ Entomol 85:2452–2458

Harvey TL, Hackerott HL (1969) Recognition of a greenbug biotype injurious to sorghum. J Econ Entomol 62:776–779

Harvey TL, Martin TJ (1988) Relative cold tolerance of Russian wheat aphid and biotype-E greenbug (Homoptera: Aphididae). J Kans Entomol Soc 61:137–140

Harvey TL, Wilde GE, Kofoid KD, Bramel-Cox PJ (1994) Temperature effects on resistance to greenbug (Homoptera: Aphididae) biotype I in sorghum. J Econ Entomol 87:500–503

Heimbach U, Eggers C, Thieme T (2004) Effect of mulch on aphid populations and virus transmission in some arrable crops. In: Simon JC, Dedryver CA, Rispe C, Hulle M (eds) Aphids in a new millennium. Proceedings of the 6th international symposium on aphids, September, 2001, Rennes, France. INRA Editions, Versailles, pp 307–312

Hentz M, Nuessly G (2004) Development, longevity and fecundity of *Sipha flava* (Homoptera: Aphididae) feeding on *Sorghum bicolor*. Environ Entomol 33:546–553

Hesler LS, Berg RK (2003) Tillage impacts cereal-aphid (Homoptera: Aphididae) infestations in spring small grains. J Econ Entomol 96:1792–1797

Hodek I, Honěk A (1996) Ecology of coccinellidae. Kluwer Academic, Dordrecht, Netherlands

House GJ, Parmelee RW (1985) Comparison of soil arthropods and earthworms from conventional and no-tillage agroecosystems. Soil Tillage Res 5:351–360

Hullé M, Maurice D, Rispe C, Simon JC (1999) Clonal variability in sequences of morph production during the transition from parthenogenetic to sexual reproduction in the aphid *Rhopalosiphum padi* (Sternorrhyncha: Aphididae). Eur J Entomol 96:125–134

Hunter SJ (1909) The greenbug and its enemies. Bull Univ Kans IX:1–163

Irwin ME, Thresh JM (1988) Long-range aerial dispersal of cereal aphids as virus vectors in North America. Philos Trans R Soc Lond B Biol Sci 321:421–446

Jyoti JL, Qureshi JA, Michaud JP, Martin TJ (2006) Virulence of two Russian wheat aphid biotypes to eight wheat cultivars at two temperatures. Crop Sci 46:774–780

Ke X, Scheu S (2008) Earthworms, Collembola and residue management change wheat (*Triticum aestivum*) and herbivore pest performance (Aphidina: *Rhophalosiphum padi*). Oecologia 157:603–617

Kelm M, Klukowski Z (2000) Weather as a factor determining damage caused by oilseed rape pests. Bull OILB/SROP 23(6):119–124

Kieckhefer RW, Elliott NC (1989) Effect of fluctuating temperatures on development of immature Russian wheat aphid (Homoptera: Aphididae) and demographic statistics. J Econ Entomol 82:119–122

Kieckhefer RW, Lytle WF, Spuhler W (1974) Spring movement of cereal aphids into South Dakota. Environ Entomol 3:347–350

Kiriac I, Gruber F, Poprawski T, Halbert S, Elberson L (1990) Occurrence of sexual morphs of Russian wheat aphid, *Diuraphis noxia* (Homoptera: Aphididae), in several locations in the Soviet Union and the northwestern United States. Proc Entomol Soc Wash 92:544–547

Kirkland RL, Peries ID, Hamilton GC (1981) Differentiation and developmental rate of nymphal instars of greenbug reared on sorghum. J Kans Entomol Soc 54:743–747

Kuroli G (1984) Laboratory investigation of the ontogenesis of oat aphis (*Rhopalosiphum padi* L). Zeitschrift für Angewandte Entomologie 97:71–76

Leather SR (1980) Egg survival of the bird cherry-oat aphid, *Rhopalosiphum padi*. Entomol Exp Appl 27:96–97

Leemans R, Cramer WP (1991) The IIASA database for mean monthly values of temperature, precipitation and cloudiness of a global terrestrial grid. Research Report RR-91-18, November 1991. International Institute for Applied Systems Analyses, Laxenburg, Austria

Li H, Payne WA, Michels GJ, Rush CM (2008) Reducing plant abiotic and biotic stress: drought and attacks of greenbugs, corn leaf aphids and virus disease in dryland sorghum. Environ Exp Bot 63:305–316

Liang HB, Zhang RZ, Zhang GX, Wen YL, Wang GP (1998) Infestation levels of *Diuraphis noxia* Mordvilko response to precipitation and irrigation. Acta Entomol Sinica 41:382–388

Lowles AJ, Tatchell GM, Harrington R, Clark SJ (1996) The effect of temperature and inoculation access period on the transmission of barley yellow dwarf virus by *Rhopalosiphum padi* (L.) and *Sitobion avenae* (F.). Ann Appl Biol 128:45–53

Lucio-Zavaleta E, Smith DM, Gray SM (2001) Variation in transmission efficiency among Barley yellow dwarf virus-RMV isolates and clones of the normally inefficient aphid vector, *Rhopalosiphum padi*. Phytopathology 91:792–796

Lukášová H, Basky Z, Starý P (1999) Flight patterns of the Russian wheat aphid, *Diuraphis noxia* (Kurdj.), during its range expansion to Central Europe (Hom.: Aphididae). Anzeiger für Schädlingskunde 72:41–44

Ma G, Ma CS (2007) Behavioral responses of bird cherry-oat aphid, *Rhopalosiphum padi*, to temperature gradients. Acta Phytophylacica Sinica 34:624–630

Malinga JN, Kinyua MG, Kamau AW, Wanjama JK, Awalla JO, Pathak RS (2007) Biotypic and genetic variation within tropical populations of Russian wheat aphid, *Diuraphis noxia* (Kurdjumov) (Homoptera: Aphididae), in Kenya. J Entomol 4:350–361

Mayo ZB Jr., Starks KJ (1974) Temperature influences on alary polymorphism in *Schizaphis graminum*. Ann Entomol Soc Am 67:421–423

McVean RIK, Dixon AFG (2001) The effect of plant drought-stress on populations of the pea aphid *Acyrthosiphum pisum*. Ecol Entomol 26:440–443

Menzel A, Sparks TH, Estrella N, Koch E, Aasa A, Ahas R, Alm-Kübler K, Bissoli P, Braslavska O, Briede A, Chmielewski FM, Crepinsek Z, Curnel Y, Dahl A, Defila C, Donnelly A, Filella Y, Jatczak K, Mage F, Mestre A, Nordli Ø, Peñuelas J, Pirinen P, Remišová V, Scheifinger H, Striz M, Susnik A, Van Vliet AJH, Wielgolski FE, Zach S, Zust A (2006) European phenological response to climate change matches the warming pattern. Glob Chang Biol 12:1969–1976

Messina FJ (1993) Winter mortality of the Russian wheat aphid (Homoptera: Aphididae) on range grasses in northern Utah. Great Basin Nat 53:310–313

Michaud JP, Grant AK (2005) Suitability of pollen sources for development and reproduction of *Coleomegilla maculata* (Coleoptera: Coccinellidae) under simulated drought conditions. Biol Control 32:363–370

Michaud JP, Jyoti JL, Qureshi JA (2006) Positive correlation of fitness with group size in two biotypes of Russian wheat aphid (Homoptera: Aphididae). J Econ Entomol 99:1214–1224

Michaud JP, Qureshi JA (2005) Induction of reproductive diapause in *Hippodamia convergens* (Coleoptera: Coccinellidae) hinges on prey quality and availability. Eur J Entomol 102: 483–487

Michaud JP, Qureshi JA (2006) Reproductive diapause in *Hippodamia convergens* (Coleoptera: Coccinellidae) and its life history consequences. Biol Control 39:193–200

Michels GJ Jr., Bateman AC (1986) Larval biology of two imported predators of the greenbug, *Hippodamia variegata* Goetz and *Adalia flavomaculata* Degeer under constant temperatures. Southwest Entomol 11:23–30

Michels GJ Jr., Behle RW (1989) Influence of temperature on reproduction, development, and intrinsic rate of increase of Russian wheat aphid, greenbug, and bird cherry-oat aphid (Homoptera: Aphididae). J Econ Entomol 82:439–444

Michels GJ Jr., Piccinnie G, Rush CM, Fritts DA (1999) Using infrared transducers to sense greenbug (Homoptera: Aphididae) infestations in winter wheat. Southwest Entomol 24:269–279

Miller RH, Pike KS, Tanigoshi LK, Buschman LL (1992) The Russian wheat aphid, *Diuraphis noxia* Mordvilko (Homoptera: Aphididae) and its natural enemies in Morocco, Jordan, Syria and Turkey. Proceedings of the 2nd Turk. Nat Con Entomol 61:68

Mittler TE, Gorder NKN (1991) Variation between clones of *Schizaphis graminum* (Homoptera: Aphididae) in the photoperiodic induction of sexual morphs. Environ Entomol 20:433–440

Newman JA (2006) Using the output from global circulation models to predict changes in the distribution and abundance of cereal aphids in Canada: a mechanistic modeling approach. Glob Chang Biol 12:1634–1642

Ogenga-Latigo MW, Baliddawa CW, Ampofo JKO (1992) Influence of maize row spacing on infestation and damage of intercropped beans by the bean aphid (*Aphis fabae* Scop). II. Reduction in bean yields. Field Crops Res 30:123–130

Olfert O, Johnson GD, Brandt SA, Thomas AG (2002) Use of arthropod diversity and abundance to evaluate cropping systems. Agron J 94:210–216

Peairs FB, Bean B, Gossen BD (2005) Pest management implications of reduced fallow periods in dryland cropping systems in the Great Plains. Agron J 97:373–377

Pierre JS (1987) Investigation of explanatory climatic variables with a view to forecasting outbreaks of insects. Utilization of a delayed integral correlation method. Bull SROP 10:109–118

Pons X, Tatchell GM (1995) Drought stress and cereal aphid performance. Ann Appl Biol 126: 19–31

Puterka GJ, Peters DC (1995) Genetics of greenbug (Homoptera: Aphididae) virulence to resistance in sorghum. J Econ Entomol 88:421–429

Rice ME, Wilde GE (1991) Aphid predators associated with conventional and conservation-tillage winter wheat. J Kans Entomol Soc 64:245–250

Rider SD Jr., Wilde GE (1998) Variation in fecundity and sexual morph production among insecticide-resistant clones of the aphid *Schizaphis graminum* (Homoptera: Aphididae). J Econ Entomol 91:388–391

Riedell WE (1989) Effects of Russian wheat aphid infestation on barley plant response to drought stress. Physiol Plant 77:587–592

Rispe C, Bonhomme J, Simon JC (1999) Extreme life cycle and sex ratio variation among sexually produced clones of the aphid *Rhopalosiphum padi* (Homoptera: Aphididae). Oikos 86:254–264

Rogers CE, Eikenbary RD, Starks KJ (1972) A review of greenbug outbreaks and climatological deviations in Oklahoma. Environ Entomol 1:664–668

Roza-Gomes MF, Salvadori JR, Schons J (2008) Damage of *Rhopalosiphum padi* (L.) (Hemiptera: Aphididae) on wheat plants related to duration time and density of infestation. Neotrop Entomol 37:577–581

Ruszkowska M (1999) Aphids as an indicator of climatic changes. Ochrona Roslin 43:9–12

Ruszkowska M (2002) Aphids on winter cereals. Ochrona Roslin 46:11–12

Ruszkowska M (2005) Protecting winter cereals against aphids, virus vectors. Ochrona Roslin 50:24–26

Ruszkowska M (2006) Permanent and cyclical parthenogenesis of *Rhopalosiphum padi* (L.) (Homoptera: Aphidoidea) across different climate regions in Poland. Prog Plant Prot 46:276–283

Sadeghi SE, Dedryver CA (2004) Effects of temperature and two isolates of barley yellow dwarf virus on transmission efficacy of *Rhopalosiphum padi* L. clones. Iran J Plant Pathol 40:Pe345, en83

Salas ML, Corcuera LJ (1991) Effect of environment on gramine content in barley leaves and susceptibility to the aphid *Schizaphis graminum*. Phytochemistry 30:3237–3240

Schweissing FC, Wilde GE (1979) Temperature and plant nutrient effects on resistance of seedling sorghum to the greenbug. J Econ Entomol 72:20–23

Sempruch C, Starczewski J, Tkaczuk A (2007) Influence of winter triticale tillage system on abundance of cereal aphids. Prog Plant Prot 47:367–370

Shufran KA, Peters DC, Webster JA (1997) Generation of clonal diversity by sexual reproduction in the greenbug, *Schizaphis graminum*. Insect Mol Biol 6:203–209

Shufran KA, Wilde GE (1994) Colanl diversity in overwintering populations of *Schizaphis graminum* (Homoptera: Aphididae). Bull Entomol Res 84:105–114

Smith JG (1969) Some effects of crop background on populations of aphids and their natural enemies on brussels sprouts. Ann Appl Biol 63:326–329

Smith JG (1976) Influence of crop background on aphids and other phytophagous insects on brussels sprouts. Ann Appl Biol 83:1–13

Smyrnioudis IN, Harrington R, Katis N, Clark SJ (2000) The effect of drought stress and temperature on spread of barley yellow dwarf virus (BYDV). Agric For Entomol 2:161–166

Sparks TH, Dennis RLH, Croxton PJ, Cade M (2007) Increased migration of Lepidoptera linked to climate change. Eur J Entomol 104:139–144

Starks KJ, Mayo ZB Jr (1985) Biology and control of the greenbug attacking sorghum. In: Leuschner K, Teetes GL (eds) Proceedings of the international sorghum entomology workshop, College Station, Texas, July 15–24, 1974, ICRISAT, USA, pp 149–158

Starks KJ, Wood EA, Teetes GL (1973) Effects of temperature on the preference of two greenbug biotypes for sorghum selections. Environ Entomol 2:351–354

Starý P (1996) Life cycle and further evidence of distribution of the Russian wheat aphid, *Diuraphis noxia* (Mordv.), in the Czech Republic. Ochrana Rostlin 32:27–31

Starý P, Lukášová H (2002) Increase of Russian wheat aphid, *Diuraphis noxia* (Kurdj.) in hot and dry weather. Anzeiger für Schadlingskunde 75:6–10

Stoetzel MB, Hammon RW (1992) New collections of sexuales of *Diuraphis* (Homoptera: Aphididae) in North America. Proc Entomol Soc Wash 94:598–599

Sumner LC, Eikenbary RD, Johnson RC (1986) Survival and reproduction of *Rhopalosiphum maidis* (Fitch) (Homoptera: Aphididae) on winter wheat during simulated drought stress. J Kans Entomol Soc 59:561–563

Susidko PI, Skylar VI (1974) Factors regulating the numbers of greenbugs on sorghum. Zashchita Rastenii 10:26

Thindwa HP, Teetes GL (1991) Effect of temperature and photoperiod on sorghum resistance to biotype C and E greenbug (Homoptera: Aphididae). J Econ Entomol 87:1366–1372

Thomas JB, Butts RA (1990) Effect of Russian wheat aphid on cold hardiness and winterkill of overwintering wheat. Can J Plant Sci 70:1033–1041

Ullah F, Peters DC (1996) Sexual reproduction capabilities of greenbugs (Homoptera: Aphididae). J Kans Entomol Soc 69:153–159

Voegtlin DJ, Halbert SE (1998) Variable morph production by some North American clones of *Rhopalosiphum padi* in response to reduced photoperiod and temperature. In: Nieto-Nafria JM, Dixon AFG (eds) Aphids in natural and managed ecosystems. Proceedings of the 5th international symposium on aphids, September 15–19, 1997, Leon, Spain, pp 309–315

Wallin JR, Loonan DV (1977) Climatic factors affecting the development of high populations of the greenbug in Texas and Oklahoma. Plant Dis Rep 61:4–6

Ward SA, Leather SR, Dixon AFG (1984) Temperature prediction and the timing of sex in aphids. Oecologia 62:230–233

Webster JA, bu Toit F, Popham TW (1993) Fecundity comparisons of the Russian wheat aphid (Homoptera: Aphididae) in Bethlehem, South Africa and in Stillwater, Oklahoma. J Econ Entomol 86:544–548

Wiktelius S (1987) Distribution of *Rhopalosiphum padi* on spring barley plants. Ann Appl Biol 110:1–7

Wraight SP, Poprawski TJ, Meyer WL, Peairs FB (1993) Natural enemies of Russian wheat aphid (Homoptera: Aphididae) and associated cereal aphid species in spring-planted wheat and barley in Colorado. Environ Entomol 22:1383–1391

Zhang RZ, Liang HB, Ren L, Zhang GX (2001) Induced life cycle transition from holocycly to anholocycly of the Russian wheat aphid (Homoptera: Aphididae). Sci China C Life Sci 44:1–7

Chapter 6
Implications of Climate Change for *Toxoptera citricida* (Kirkaldy), a Disease Vector of Citrus in Florida

Jawwad A. Qureshi

Abstract Increasing temperatures, elevated CO_2 levels, and changes in rainfall patterns are predicted to impact plants and insects, both harmful and beneficial. *Toxoptera citricida* Kirkaldy (Homoptera: Aphididae), commonly known as the brown citrus aphid (BrCA), is a cosmopolitan pest of citrus and a highly efficient vector of citrus tristeza virus (CTV). Both the pest and the disease pose a serious threat to citrus production in Florida. Temperature is the most important abiotic factor impacting the biology of BrCA and the growth of new citrus shoots on which the aphid depends for development and reproduction. Climate in most parts of the state is humid subtropical, characterized by mild to cool, relatively dry winters and autumns and hot, wet springs and summers. Cool winters, freezes and hot summers limit the aphid's ability to survive and reproduce, leaving spring as the optimal season for population growth. Mature citrus trees go through a period of winter dormancy followed by a surge of growth in spring and periods of sporadic growth in summer and fall. Any significant rise in average winter temperatures will reduce the severity and duration of cool winter weather and may trigger earlier shoot growth by citrus trees. This would improve aphid survival in winter, provide a food bridge between fall and spring populations, and possibly lead to earlier recruitment of natural enemies, particularly predators. Summers, on the other hand, if warmer and more prolonged would probably be less tolerable for the BrCA and its natural enemies. However, adaptation to environmental change is possible and low aphid populations are observed in the field at temperatures above the upper threshold for development determined under laboratory conditions. The negative impact of rising temperatures on BrCA populations may be more pronounced in the south than elsewhere due to the relatively higher temperatures there, with more beneficial effects evident in the north.

Keywords Brown citrus aphid · Citrus trees · *Toxoptera citricida* · Tristeza virus

---

J.A. Qureshi (✉)
Department of Entomology and Nematology, Institute of Food and Agricultural Sciences, University of Florida, Immokalee, FL 34142, USA
e-mail: jawwadq@ufl.edu

P. Kindlmann et al. (eds.), *Aphid Biodiversity under Environmental Change*,
DOI 10.1007/978-90-481-8601-3_6, © Springer Science+Business Media B.V. 2010

6.1 Introduction

Climate plays a very important role in citrus production through its impacts on the plant itself and related insect pests and diseases. Citrus is produced in tropical and subtropical regions around the world due to its lack of cold tolerance. Most production is concentrated in subtropical areas because a period of tree dormancy induced by cool winter temperatures helps to enhance fruit quality during ripening (Garcia-Luis et al. 1992; Rosenzweig et al. 1996). However, there is always a serious risk of freeze in these areas which damages newly developing shoots and can cause significant die-back and even tree death (Turrell 1972; Wiltbank and Oswalt 1987; Huang et al. 1993). Favorable temperatures for plant growth range from 12.8 to 37.8°C during the growing season and 1.7–10°C during winter dormancy. Although mature trees may survive up to 10 h below −3.9°C, most fruit is damaged at these temperatures and very young trees are killed within a range of 1.1–3.3°C. Excessively warm temperatures during periods of bloom or early fruit set can induce fruit abscission (Moss 1969; Reuther 1973; Ono et al. 1988) but also, during summer months, reduce pest populations, particularly aphids. Preferable annual precipitation ranges from 12.5 to 50 cm, although oranges are frequently grown in areas receiving 100–150 cm.

Florida citrus accounts for 9% of the nine million acres of current citrus acreage worldwide and yields 15% of the world's production. Florida produces 20% of the world's oranges on about 10% of the acreage, and approximately half of the world's grapefruit on about half of the total acreage. Citrus is the most important agricultural crop in Florida with an annual value of $1.5 billion (Anonymous 2008), about 4–6% of the orange crop, and as much as 35–40% of the grapefruit crop, sold as fresh fruit. Most citrus is grown in the central part of the state, mainly because production in the southern part is limited by poorly drained soils and, in the north, by higher freeze probabilities. Arctic air masses that bring freezing temperatures to Florida are typically moderated somewhat as they traverse the length of the state and the northern limits of commercial citrus production have been established by the history of damaging freezes. Even in southwest Florida, some recent freezes have caused serious damage to foliage, fruit and insect pests. In February 2007, temperatures dropped to 0°C for two nights and killed more than 50% of new shoots on 6 year old "Valencia" orange trees along with the immature aphids and psyllids infesting them (Qureshi and Stansly 2009).

New citrus growth is attacked by three serious insect pests, two of which are vectors of devastating citrus diseases. *Diaphorina citri* Kuwayama (Hemiptera: Psyllidae), commonly known as the Asian citrus psyllid (ACP), vectors *Candidatus* Liberibacter asiaticus, a bacterium that causes "huanglongbing" or citrus greening disease (Halbert and Manjunath 2004). *Toxoptera citricida* Kirkaldy (Homoptera: Aphididae), commonly known as brown citrus aphid (BrCA), is the most important vector of citrus tristeza virus (CTV) (Michaud 1998). *Phyllocnistis citrella* Stainton (Lepidoptera: Gracillariidae), commonly known as citrus leafminer (CLM), can cause substantial direct damage to new foliage and also exacerbates infections of citrus canker, a disease caused by the bacterium *Xanthomonas axonopodis pv. citri* Hasse (Gottwald et al. 1997).

Brown citrus aphid is an economically important pest of citrus mainly due to the high efficiency with which it transmits CTV. CTV is a phloem-limited closterovirus that can cause devastating losses in citrus, particularly in regions where sour orange has been relied on as a rootstock. Although some variants of CTV are benign, others cause "quick decline" of trees on sour orange rootstock and "stem-pitting" symptoms irrespective of variety. Although much CTV has been spread through propagation of infected cuttings, epidemics of the disease have been precipitated by invasion of BrCA in Venezuela, Brazil, Colombia, Uruguay, Argentina, and Peru (Rocha-Peña et al. 1995). BrCA is indigenous to Southeast Asia (Kirkaldy 1907; Tao and Tan 1961; Carver 1978; Rocha-Peña et al. 1995) but has now spread to most citrus-producing regions of the world, with the exception of the Mediterranean (Michaud 1998). In the United States, BrCA was first identified in Ft. Lauderdale, Florida in 1995 and, within 1 year, spread as far north as Melbourne on the east coast and Ft. Meyers in the southwest (Halbert 1997). By the summer of 1997, BrCA was widely distributed throughout south-central and coastal regions of Florida, although it remained absent from much of the ridge citrus north of Highlands county (Michaud 1998). In 2000, BrCA was discovered in the Yucatan peninsula (Michaud and Alvarez 2000) and by 2002, it was distributed throughout all citrus-producing regions of Florida (Liu and Tsai 2002). BrCA is the primary vector of CTV throughout Asia, Africa, Australia, South America, Central America and Florida (Halbert and Brown 1996; Michaud 1998). Although CTV was present in Florida before BrCA was identified, the rate of spread of the disease increased dramatically after invasion of the vector (Tsai et al. 2000).

Climate systems are complex and current predictions of the positive and negative effects of climatic change on flora and fauna are admittedly imprecise. Nevertheless, there is scientific consensus that natural factors and human activities will continue to influence climate and result in increasing temperatures, elevated levels of CO_2, changes in precipitation, and changes in prevailing wind patterns. The cascading effects of changes in temperature and other climatic variables will impact the nutritional status and suitability of host plants for insect herbivores, the developmental and reproductive rates of pests and their natural enemies, and overall pest management strategies (Storey and Storey 1986; Stone and Willmer 1989; Yocum et al. 1991). Temperature is the single most important abiotic factor affecting the development and reproduction of aphids, although geographically distinct aphid populations may differ in their tolerance of temperature extremes (Campbell et al. 1974). In this chapter I discuss potential implications of climate change for BrCA populations in Florida citrus.

6.2 Florida's Climate

The peninsula of Florida is bordered by the Gulf of Mexico on the west and the Atlantic Ocean on the east, with the Caribbean Sea located just off the southern tip. Therefore, its climate is tempered by the bodies of water surrounding it, in particular the warm waters of the Gulf Stream. Florida's climate is perhaps better characterized by precipitation than by temperature, with mild to cool, relatively dry winters and autumns (the dry season) and hot, wet springs and summers (the wet season). The climate in most parts of the state is humid subtropical, except for the south which borders on tropical, and the Florida Keys which are fully tropical (http://www.virtualamericas.net/usa/photos/coast/floridakeys/us0193.shtml). Cold fronts during late fall and winter can occasionally bring high winds and cool to cold temperatures to the entire state. The Gulf Stream has a moderating effect on the climate, and although much of Florida commonly experiences summer temperatures exceeding 32°C, they seldom exceed 39°C. The hottest temperature recorded in the state was 43°C on June 29, 1931 in Monticello and the coldest was −19°C on February 13, 1899, just 25 miles away, in Tallahassee. Mean high temperatures during summer are generally in the range of 30–35°C. Mean low temperatures in winter generally range from 4 to 10°C and average 2–3°C higher in southern Florida. Severe weather is common from late spring until early autumn, with frequent thunderstorms converging on the peninsula from both eastern and western coasts. Consequently, Florida has the highest average precipitation of any state in the USA. Florida also leads the United States in the frequency of tornadoes per square mile, but these typically do not reach the size or intensity of those in the Midwest and Great Plains.

The average temperature in Florida has increased over the past century (EPA 230-F-97–008i 1997). At Ocala, the 1892–1921 average temperature was almost 19°C, whereas from 1966 to 1995 it was over 21°C. Significant changes in precipitation patterns are one likely consequence of further increases in mean temperatures. Sea level rise is another consequence and some coastal regions will be more adversely impacted than others (Yin et al. 2009). The highest land in Florida is barely 100 m ASL and yet, along much of the coast, sea level is rising by 7–9 inches per century and is projected to rise 18–20 inches by 2100 (EPA 230-F-97–008i 1997). Precipitation over the last 100 years has decreased in the Keys and parts of south Florida, but increased in central Florida and along the panhandle (Karl et al. 1996). A significant reduction in rainfall was observed during winter, spring and early summer from 2005 to 2007 compared to the 30-year average at Immokalee, Florida (Fig. 6.1, Qureshi and Stansly 2008). An anomalous set of damaging freezes during several winters in the 1980s reinforced speculation that a cooling trend was apparent in the record since the mid 1950s (Chen 1984, 1985). Some projected that the cooling trend would continue or had bottomed out (Winterling 1984), but others thought these cold-temperature events occurred by chance and without any change in the underlying probability distribution (Glantz and Katz 1987).

Over the next century, more climatic changes are expected in Florida. Data from Intergovernmental Panel on Climate Change and United Kingdom Hadley Centre's climate model (HadCM2) indicate that, by 2100, mean temperatures in Florida could increase by 0.09–0.12°C (range: 0.03–0.18°C) in spring, summer, and fall, and by somewhat less in winter (EPA 230-F-97–008i 1997). A recent climate report produced by 13 federal agencies and several major universities and research centers reported that mean temperatures in the mainland US would increase anywhere from 0.21 to 0.34°C by 2090, with some margin of error. These predictions are similar to

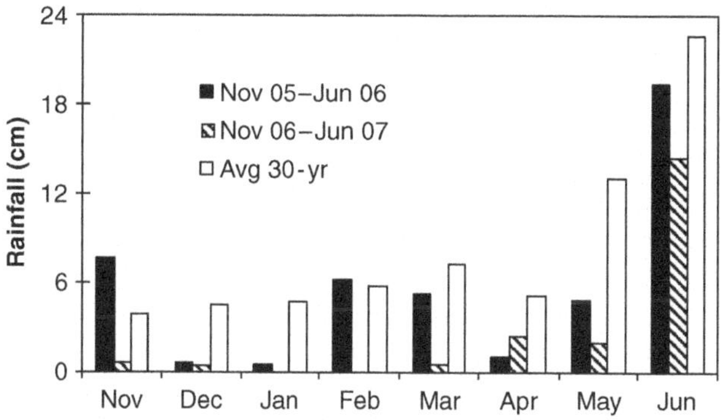

Fig. 6.1 Monthly rainfall between November and June from 2005 to 2007 and past 30 years at Southwest Florida Research and Education Center, University of Florida, Immokalee, Florida, USA

the ones from 2007 report by the UN's Intergovernmental Panel on Climate Change. The frequency of extremely hot days in summer is expected to increase. For example, Southern Florida could see more than 160 days a year above 32°C. Precipitation is generally predicted to increase in north and decrease in the south, especially the southwest.

6.3 Citrus Production

Only a small proportion of world's agricultural land is suitable for citrus production due to requirements for seasonal and diurnal temperature differentials and adequate precipitation for the production of high quality fruit. Citrus has been an important industry in Florida since well before the turn of the century specifically because of the suitable climate, but there have always been threats from severe weather events and citrus pests and diseases have colonized the state over time. The severe freezes of the 1980s are estimated to have killed one third of state's commercial citrus trees (Miller and Glantz 1988; Miller 1991) and contributed to a general shift in citrus production southward. More recently, Florida has lost millions of citrus trees to diseases such as citrus canker, huanglongbing, and CTV. Now, climate change stands to change the dynamics of citrus diseases and disease vectors such as BrCA by direct effects on the life history of these organisms, and by indirect effects on host plants and natural enemies.

Brown citrus aphids require newly expanded citrus shoots (flush) to grow and reproduce. These remain suitable for growth and reproduction for a period of 3–4 weeks, depending on environmental conditions (Michaud 1998). Seasonal patterns of flushing vary and are influenced by weather, tree age, and citrus variety (Knapp et al. 1995). Flush production occurs earlier, more frequently, and more vigorously in young trees than in mature ones. In Florida, most flush is produced during early spring, relatively less in early summer, and some during late summer and fall, with no significant growth occurring during the dormant winter period (Cooper et al. 1963; Hall and Albrigo 2007; Qureshi et al. 2009). Any significant increase in winter temperatures could trigger earlier onset of shoot production and earlier aphid infestions. Hall et al. (2008) found a positive correlation between air temperatures and shoot production in mature orange trees in east-central Florida. They observed that the typical peak of spring growth during March was absent in 2006, an occurrence they linked to a lack of rainfall. Comparing December 2007 with December 2006 in Immokalee, FL, average and maximum temperatures were 1°C higher and rainfall was 1 cm greater, which resulted in significant early flush production by mature trees in January 2008, rather than in late February or early March (JA Qureshi, unpublished data). Thus, one predicted effect of a warming climate is a shift in the seasonal availability of food for BrCA to earlier in spring.

6.4 Climate Change and Brown Citrus Aphid

Aphids, like all insects, are poikilothermic and have a large surface to volume ratio compared to larger animals, making temperature regulation a challenge. Temperature is probably the single most important environmental factor impacting aphid biology, ecology and behavior, largely overwhelming the effects of other environmental factors (Dixon 2000; Bale et al. 2002). Although Newman (2006) found no common response among cereal aphids to several climate changes projected using a mechanistic model, temperature was the only variable projected to impact multiple aphid responses. It has been estimated that a 2°C increase in annual mean temperature could potentially result in one to five additional aphid generations per season, depending on species (Yamamura and Kiritani 1998). However, negative impacts of high temperatures on aphid survival and reproduction have often been noted. For example, Ma et al. (2004) predicted that a 1°C increase in average daily temperature and 1.3°C increase in maximum daily temperature during the period of aphid population increase would result in a 33% reduction in peak abundance of rose grain aphid, *Metopolophium dirhodum*, whereas an increase of 2°C in average and 2.6°C in maximum daily temperature would result in a 74% reduction and earlier population collapse. Consequently, global warming could greatly decrease aphid infestations of spring-sown cereal crops grown throughout Germany and Central Europe.

Tsai and Wang (1999) evaluated development, survivorship, and reproduction of BrCA at eight constant temperatures. The developmental period ranged from 63.1 d at 8°C to 5.5 d at 30°C and the lower developmental threshold was estimated to be 6.3°C. Survivorship of nymphs varied from 81 to 97% within a temperature range of 8–30°C and was reduced to 29% at 32°C. Adult female longevity ranged from 60.0 d at 10°C to 6.5 d at 32°C. Lifetime fecundity ranged from 52.5 nymphs per female at 20°C to 7.5 at 32°C. The largest intrinsic rate of increase ($r_m = 0.3765$) occurred at 28°C, compared with the smallest r_m values of 0.0588 and 0.0960 for populations reared at 10 and 32°C, respectively. The mean generation time ranged from 51 d at 10°C to 8 d at 32°C. Tang et al. (1999) examined the influence of temperature on development and population growth of BrCA on sour orange seedlings at 20, 25, and 30°C with 60–80% RH and a photoperiod of 14:10 (L:D). Aphid development at 20°C was ≈1.5 d longer than that at 25°C and 2.7 d longer than that at 30°C. The reproductive period was 22.2 d at 20°C, 7 d and 11 d longer than at 25 and 30°C, respectively. However, fecundity at 20°C was 20% greater than that at 25°C, and 80% greater than that at 30°C. Tang et al. (1999) concluded that 25°C was the most suitable temperature for BrCA population growth because it resulted in the highest r_m and shortest doubling time (DT). Although the reproductive rate of individuals was highest at 20°C, both r_m and DT were sub-optimal at this temperature. Earlier observations by Komazaki (1982) in Japan yielded similar results, with highest r_m at a constant temperature of 27°C, even though the fecundity and net reproductive rate (R_0) of apterous females was maximal at 21.5°C. However, none of the above studies examined the production of alatae in BrCA colonies and these play a vital role in the spread of aphid infestations and CTV transmission. Most transmission of CTV in the field is thought to occur when alates leave maturing colonies on infected trees and colonize uninfected ones. Michaud (2001a) found that the proportion of BrCA maturing into alatae was positively correlated with colony density. Takanashi (1989) found that pre-reproductive period was longer for alatae than for apterae at both 20 and 25°C, and that fecundity and R_0 were lower at both temperatures, consistent with most other aphid species.

Laboratory experiments typically examine effects of constant temperatures, whereas diurnal cycles are more characteristic of natural conditions. Liu et al. (1995) concluded that, for phenological models, differences between constant and varying temperature regimes with the same mean can be ignored. In addition, heat transfer in air columns under natural field conditions may be quite different from the "still air" conditions of laboratory growth chambers. Since BrCA is very sensitive to temperature changes near both the lower and upper thresholds, any significant increase in average winter temperatures will improve aphid survival, whereas higher than optimal temperatures in summer will reduce reproduction and nymphal survival, as was observed in the field by Michaud and Belliure (2000).

6.4.1 Fall and Winter Populations of BrCA

Sporadic fall flushes present the last opportunity for BrCA population to increase before temperatures become too cold for citrus growth. Citrus scions remain largely dormant during winter months on all but the youngest of trees, but there are always one or two tender shoots available. Thus, BrCA is able to survive on "resets" (young trees planted as replacements) and isolated root sprouts that invariably occur somewhere in every mature grove. Tsai and Wang (1999) reported that 88% of nymphs survived to adulthood at 8°C, but none matured at 5°C, implying a developmental threshold range of 5–8°C. Nevertheless, short exposures to temperatures between

1 and 5°C are unlikely to be fatal to aphid colonies. Minimum temperatures in Florida generally range between 4–10°C during winter but sometimes reach 0°C during cold fronts and freezes. Abscision of new shoots as a consequence of freezing also kills all insects feeding on them. For example, significant mortality to immature of *D. citri* on young shoots of 6 year old "Valencia" orange trees was observed during a freeze in February 2007 (Qureshi and Stansly 2009). In addition, since the BrCA lacks a holocycle in the new world, it does not produce eggs that are capable of withstanding freezing temperatures. It can be inferred that any significant increase in winter temperature minima would result in more winter flush and improve BrCA overwintering survival, possibly leading to more widespread outbreaks in spring. However, according to the Hadley Centre's climate model (HadCM2) temperature increases in winter will be somewhat less compared to those in spring, summer, and fall. Nevertheless, warming would be expected to shift the spring population peak earlier in the year, and extend aphid activity later in fall, thus reducing the period aphids are exposed to adverse winter conditions. The net effect on aphid populations and disease transmission will depend largely on the extent to which generalist predators are able to track these changes in aphid population cycles.

6.4.2 Spring and Summer Populations of BrCA

Cooper et al. (1963) reported that from 76 to 87% of the annual production of new shoots by mature "Valencia" orange trees in Florida occurred during February and March. In addition to reducing the duration and severity of winter conditions, a significant rise in temperature would advance the onset of shoot production in spring, favoring earlier development of aphid infestations. Summer temperatures in Florida generally reach and exceed 30°C on a daily basis, but seldom exceed 40°C. Temperatures beyond 28°C negatively impact BrCA reproduction and nymphal survival. Tsai and Wang (1999) reported that, at 32°C, females lived for only 6.5 days and produced an average of only 7.5 nymphs with a survival rate of only 29%. The r_m value of 0.0960 obtained at 32°C was three and four times lower than values obtained at 25 and 28°C, respectively (Tang et al. 1999; Tsai and Wang 1999). Thus, high temperatures are largely responsible for the low populations of BrCA observed during summer even when suitable flush is available. However, small colonies can survive high ambient temperatures in protected microclimates. Michaud (1999b) followed the process of colony formation and maturation in various cohorts of BrCA in Florida citrus groves, some during summer when temperatures frequently exceeded optimal values. For example, two cohorts followed in Ft. Pierce persisted at temperatures averaging 27.9°C for the study period, with daily maxima ranging from 31.1 to 34.4°C. Similarly a June cohort followed in Immokalee experienced mean temperatures averaging 27.6°C and maxima ranging from 31.7 to 35.0°C. The author speculated that some colony mortality in the unknown category was due to high temperatures. Another study conducted in a citrus orchard in Lake Alfred,

Florida provided further evidence of the negative impact of heat on BrCA colonies (Fig. 6.2; Michaud and Belliure 2000). When daily temperatures averaged 29.3 ± 0.3°C and maximum temperatures regularly exceeded 35°C for several hours, longevity and fecundity of apterous adults was reduced by 50 and 80%, respectively, and nymphal mortality approached 100%. These findings are supportive of laboratory studies that indicate temperatures beyond 28°C will negatively impact

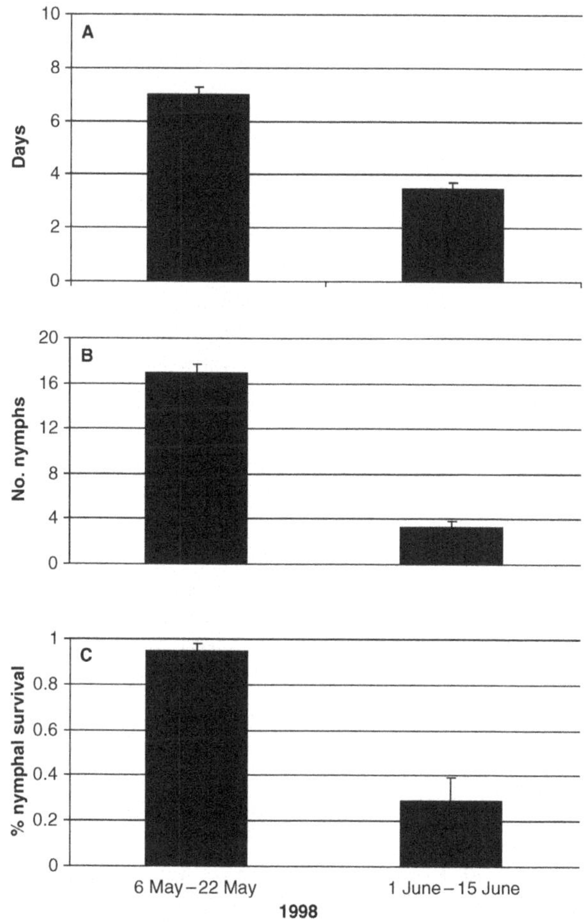

Fig. 6.2 Evidence of negative impact of high temperature on apterous *Toxoptera citricida* longevity (**a**), fecundity (**b**) and progeny survival (**c**) on potted Carrizo citrange seedlings interspersed between established trees in a 5-year old grove of Orlando tangelo and manually infested with either apterous or alate adult T. citricida at Lake Alfred, Florida, USA. Averages of mean and maximum daily temperatures were 25.5±0.3°C and 31.1±0.3°C, respectively, for 6–22 May, and 29.3±0.3°C and 36.0±0.4°C, respectively, for 1–15 June

BrCA growth and reproduction. Insects show a sharp rise in activity once ambient temperature exceeds the preferred range, followed by torpor and death if high temperatures persist. Since insects can utilize evaporative cooling to withstand high temperatures in dry air, high humidity exacerbates the impact of heat. For example, the American cockroach, *Periplaneta americana* (L.), dies at 38°C in high humidity, but can survive short exposures of up to 48°C in dry air (Chapman 1998). Although increasing summer temperatures are likely to drastically reduce BrCA populations in Florida, Tang et al. (1999) found that Florida BrCA populations were able to tolerate higher temperatures than those in Japan. They found that fecundity of BrCA reared at 30°C in Florida was ≈10 times greater, and r_m twice as large, compared to Japanese strains reared under the same conditions. The existence of variation among BrCA populations in thermal tolerance raises the question of whether the Florida population could evolve even greater tolerance to high temperature over time.

Increases in atmospheric CO_2 concentrations and changes in precipitation patterns are also likely to alter plant phenology and affect the growth and abundance of herbivores and their natural enemies (Hamilton et al. 2005; Coviella and Trumble 1999; Hunter 2001). Elevated levels of CO_2 either diminish or improve plant quality for aphids which, in turn, could alter their rates of consumption and growth. Some studies predict increased fecundity and abundance of sucking pests as a consequence of increased CO_2 (Bezemer et al. 1998), which could result in shorter generation times and aphid population growth rates that more often exceed rates of aphid removal by predators. Although Michaud (1999b) found that few BrCA colonies were able to escape discovery by predators, faster growing colonies are more resilient to predation by virtue of reaching large size sooner. Only a fraction of colonies become large enough to produce significant numbers of alates and these are critical to the spread of CTV. Predators and parasitoids may also respond positively to elevated CO_2 levels and become more abundant and efficacious (Stilling et al. 1999; Percy et al. 2002). *Harmonia axyridis* Pallas (Coleoptera: Coccinellidae), consumed more *Aphis gossypii* Glover (Hemiptera: Aphididae), under conditions of higher CO_2 (Chen et al. 2005). However, other experiments (Chen et al. 2007) revealed no significant effects of elevated CO_2 on rates of *H. axyridis* growth and development on *Sitobion avenae* F. (Hemiptera: Aphididae), but increased abundance of the parasitoid, *Aphidius picipes* (Nees) (Hymenoptera: Braconidae).

Generally, there are two distinct population peaks of BrCA annually, one each in the spring and autumn (Chagas et al. 1982; Nickel and Klingauf 1985; Seif and Islam 1988; Michaud and Browning 1999; Tsai and Lee 1999). However, the peaks do not fall on the same months each year and are solely dependent on rainfall in the preceding month that leads to the production of new shoots. Changes in rainfall patterns could alter flush cycles, thus affecting food availability for BrCA populations. Although heavy rains have been implicated in dislodging aphids from plants (Dunn and Wright 1955), Michaud (1999b) counted aphids in BrCA colonies before and after torrential downpours (ca. 31 cm total) in Puerto Rico and found that the majority of aphids remained unaffected, likely because they feed on abaxial leaf surfaces with their stylets deeply embedded in leaf tissues.

6.4.3 Biological Control of BrCA

Control of BrCA is critical to reduce the spread of CTV. Natural enemies such as predators, and to a lesser extent parasitoids and entomopathogenic fungi, play an important role in the management of BrCA in Florida and will also respond to climate change. A significant increase in temperatures during winter and spring will likely result in improved environmental conditions for natural enemies with increased abundance of prey and alternative foods such as honey dew and pollen. Improved reproductive performance of natural enemies earlier in the season could improve biological control or at least maintain it at present levels. However, the negative impact of high temperatures in summer will probably be greater for natural enemies than for their aphid prey, largely because of the higher activity levels associated with foraging behavior. Since summer is the most challenging season for most beneficial insects in Florida citrus, a lengthening of the season would not be likely to improve levels of biological aphid control.

Parasitism of BrCA in Florida is usually only a minor source of mortality. Only the native *Lysiphlebus testaceipes* (Cresson) (Hymenoptera: Aphidiidae) sometimes contributes appreciable mortality of BrCA. Weathersbee et al. (2004) concluded that the high thermal threshold and low degree-day requirement for development of *L. testaceipes* favored its persistence under Florida conditions. However, only 7% colonies suffered from parasitism in the spring cohorts followed by Michaud (1999a) and parasitism was never a cause of colony extinction and any appreciable increase in summer temperatures is likely to negatively impact reproduction and survival of the parasitoids. For example, Liu and Tsai (2002) observed a 60% reduction in reproduction of *Lysiphlebia mirzai* (ShujaUddin), another parasitoid of BrCA, when temperature increased from 25 to 32°C for 24 h.

Various generalist predators, particularly coccinellids, syrphids, and chrysopids, attack the BrCA in Florida (Michaud 1999b, 2000, 2001b; Belliure and Michaud 2001). BrCA were acceptable prey to seven species of ladybeetles, however, only *Cycloneda sanguinea* (L.) and *H. axyridis* completed development and only *C. sanguinea* laid eggs on an exclusive BrCA diet (Michaud 1999b). Field studies have shown that predator populations, particularly ladybeetles, track the abundance of citrus aphids and are high in spring when temperatures are favorable and low in summer and winter when temperatures negatively impact both BrCA and its natural enemies (Michaud 1998; Qureshi and Stansly 2009). Whereas a rise in winter temperatures could extend the activity of predators later in the fall and lead to earlier activity in spring, it is not clear whether this would compensate for longer periods of forced inactivity in summer. However, ladybeetles may survive in other agro-ecosystems and return to colonize citrus when it becomes infested with suitable prey. For example, *Curinus coeruleus* Mulsant, *Olla v-nigrum* Mulsant, *H. axyridis*, and *C. sanguinea* were observed during summer on trees infested with *D. citri*, *T. citricida*, and *Aphis spiraecola* Patch (Qureshi and Stansly 2007, 2009, 2010). Thus, some species of ladybeetles may remain active in citrus groves during summer and a shift in species composition of the predator community is one possible outcome of climate change. For example, *C. coeruleus* is a tropical species of Mexican origin that tolerates high temperatures well and can be fully active in mid-summer when many other coccinellids are not. It is also quite possible that the direct negative effects of high temperatures on BrCA will be more than enough to offset those on its primary natural enemies.

6.5 Conclusions

Climate warming is generally expected to negatively impact BrCA populations in Florida, a region where high temperatures combined with high humidity are already important limiting factors to aphid population growth through much of the year. A significant increase in mean temperatures during winter and spring will induce earlier growth of new shoots and aphid infestations in citrus, in turn leading to earlier activity of aphid natural enemies. Aphid reproduction and survival could be significantly reduced in summer if there are longer periods when temperature remains above the optimal threshold for aphid growth. This effect could be enhanced if there is reduced availability of summer flush as a consequence of hot, dry conditions. Hotter, longer summers would also imply extended periods of prey scarcity for aphid natural enemies and could further impede their ability to survive this difficult season.

Acknowledgements I thank Dr. J.P. Michaud for helpful review of this chapter.

References

Anonymous (2008) Citrus summary 2006–2007. Florida Agricultural Statistics Service, Florida, Florida Department of Agriculture and Consumer Services. (http://www.nass.usda.gov/Statistics_by_State/Florida/Publications/Citrus/cs/2006–07/CS0607all.pdf)

Bale JS, Masters GJ, Hodkinson ID, Awmack C, Bezemer TM, Brown VK, Butterfield J, Buse A, Coulson JC, Farrar J, Good JEG, Harrington R, Hartley S, Jones TH, Lindroth RL, Press MC, Symrnioudis I, Watt AD, Whittaker JB (2002) Herbivory in global climate change research: direct effects of rising temperatures on insect herbivores. Glob Chang Biol 8:1–16

Belliure B, Michaud JP (2001) Biology and behavior of *Pseudodorus clavatus* (F.) (Diptera: Syrphidae), an important predator of citrus aphids. Ann Entomol Soc Am 94:91–96

Bezemer TM, Jones TR, Knight KJ (1998) Long-term effects of elevated CO_2 and temperature on populations of the peach potato aphid *Myzus persicae* and its parasitoid *Aphidius matricariae*. Oecologia 116:128–135

Campbell A, Frazer BD, Gilbert N, Gutierrez AP, Mackauer M (1974) Temperature requirements of some aphids and their parasites. J Appl Ecol 11:431–438

Carver M (1978) The black citrus aphids, *Toxoptera citricidus* (Kirkaldy) and *T. aurantii* (Boyer de Fonscolombe) (Homoptera: Aphididae). J Aust Entomol Soc 17:263–270

Chagas EFD, Neto SS, Braz AJBP, Mateus CPB, Coelho IP (1982) Population fluctuations of pest and predator insects in citrus. Pesqui Agropecu Bras 17:817–824 (in Portuguese, English summary)

Chapman RR (1998) The insects. Structure and function. Cambridge University Press, UK

Chen E (1984) Long-term climate trend. In: Cold protection guide. Lake-orange extension service, central florida freeze recovery task force. Institute of Food and Agricultural Sciences, University of Florida, Gainesville, Fla, pp 140–143

Chen E (1985) Minimum temperature change and freeze probabilities in Florida: 1925–1984. In: Cold protection guide, 1985 revision. Institute of Food and Agricultural Sciences, University of Florida, Gainesville, Fla, pp S1–S5

Chen FJ, Gang W, Megha N, Parajulee FG (2007) Impact of elevated CO_2 on the third trophic level: a predator *Harmonia axyridis* and a parasitoid *Aphidius picipes*. Biocontrol Sci Technol 17:313–324

Chen F, Ge F, Parajulee MN (2005) Impact of elevated CO_2 on tritrophic interaction of *Gossypium hirsutum*, *Aphis gossypii* and *Leis axyridis*. Environ Entomol 34:37–46

Cooper WC, Peynado A, Furr JR, Hilgeman RH, Cahoon GA, Boswell SB (1963) Tree growth and fruit quality of Valencia oranges in relation to climate. In Proceedings, American Society of Horticultural Science, 27–29 Aug. 1963, Alexandria, VA. American Society of Horticultural Science, Alexandria, VA, pp 82, 180–192

Coviella C, Trumble J (1999) Effects of elevated atmospheric carbon dioxide on insect plant interactions. Conserv Biol 13:700–712

Dixon AFG (2000) Aphid ecology. Chapman and Hall, London

Dunn JA, Wright DW (1955) Population studies of the pea aphid in East Anglia. Bull Entomol Res 46:369–387

EPA 230-F-97–008i (1997) Climate change and Florida

Garcia-Luis A, Kanduser M, Santamarina P, Guardiola JL (1992) Low temperature influence on flowering in Citrus: the separation of inductive and bud dormancy releasing effects. Physiol Plant 86:648–652

Glantz M, Katz R (1987) African drought and its impacts: revived interest in a recurrent phenomenon. Desertif Control Bull 14:23–32

Gottwald TR, Graham JH, Schubert TS (1997) An epidemiological analysis of the spread of citrus canker in urban Miami, Florida, and synergistic interaction with the Asian citrus leafminer. Fruits 52:371–378

Halbert SE (1997) Brown citrus aphid: current situation and prognosis. Cit. Ind., April Issue

Halbert SE, Brown LG (1996) *Toxoptera citricida* (Kirkaldy), brown citrus aphid – identification, biology and management strategies. Florida Department *of* Agriculture *& Consumer* Services, Division Plant Industry, Entomol. Cir. No. 374

Halbert SE, Manjunath KL (2004) Asian citrus psyllids (Sternorrhyncha: Psyllidae) and greening disease of citrus: a literature review and assessment of risk in Florida. Fla Entomol 87:330–353

Hall DG, Albrigo LG (2007) Estimating the relative abundance of flush shoots in citrus with implications on monitoring insects associated with flush. HortScience 42:364–368

Hall D, Hentz MG, Adair RC (2008) Population ecology and phenology of *Diaphorina citri* (Hemiptera: Psyllidae) in two Florida citrus groves. Environ Entomol 37:914–924

Hamilton JG, Dermody O, Aldea M, Zangerl AR, Rogers A, Berenbaum MR, DeLucia EH (2005) Anthropogenic changes in tropospheric composition increase susceptibility of soybean to insect herbivory. Environ Entomol 34:479–485

Huang S, Li G, Zhou C, Fan X, Shen C, Xue M (1993) A climatological study of injury to citrus trees from freezing weather in China. Agric For Meteorol 65:129–138

Hunter MD (2001) Effects of elevated atmospheric carbon dioxide on insect-plant interactions. Agric For Entomol 3:153–159

Karl TR, Knight RW, Easterling DR, Quayle RG (1996) Indices of climate change for the United States. Bull Am Meteorol Soc 77:279–292

Kirkaldy GW (1907) On some peregrine Aphidae in Oahu, Honolulu. Proc Hawaiian Entomol Soc 1:100

Knapp JL, Albrigo LG, Browning HW, Bullock RC, Heppner JB, Hall DG, Hoy MA, Nguyen R, Pena JE, Stansly PA (1995) Citrus leafminer, *Phyllocnistis citrella* Stainton: current status in Florida. Florida Cooperative Extension Service, Institute of Food and Agricultural Sciences, University of Florida, Gainesville, FL

Komazaki S (1982) Effects of constant temperatures on population growth of three aphid species, *Toxoptera citricida* (Kirkaldy), *Aphis citricola* van der Goot, and *Aphis gossypii* Glover (Homoptera: Aphididae) on citrus. Appl Entomol Zool 17:75–81

Liu YH, Tsai JH (2002) Effect of temperature on development, survivorship, and fecundity of *Lysiphlebia mirzai* (Hymenoptera: Aphidiidae), a parasitoid of *Toxoptera citricida* (Homoptera: Aphididae). Environ Entomol 31:418–424

Liu SS, Zhang GM, Zhu J (1995) Influence of temperature variations on rate of development in insects: analysis of case studies from entomological literature. Ann Entomol Soc Am 88: 107–119

Ma CS, Hau B, Poehling MM (2004) The effect of heat stress on the survival of the rose grain aphid, *Metopolophium dirhodum* (Hemiptera: Aphididae). Eur J Entomol 101:327–331

Michaud JP (1998) A review of the literature on *Toxoptera citricida* (Kirkaldy) (Homoptera: Aphididae). Fla Entomol 81:37–61

Michaud JP (1999a) Aggregation by alatae of *Toxoptera citricida* (Homoptera: Aphididae). Environ Entomol 28:205–211

Michaud JP (1999b) Sources of mortality in colonies of the brown citrus aphid, *Toxoptera citricida* (Homoptera: Aphididae). Biocontrol 44:347–367

Michaud JP (2000) Development and reproduction of ladybeetles (Coleoptera: Coccinellidae) on the citrus aphids *Aphis spiraecola* Patch and *Toxoptera citricida* (Kirkaldy) (Homoptera: Aphididae). Biol Control 18:287–297

Michaud JP (2001a) Colony density and wing development in *Toxoptera citricida* (Homoptera: Aphididae). Environ Entomol 30:1047–1051

Michaud JP (2001b) Evaluation of green lacewings, *Chrysoperla plorabunda* Fitch (Neuroptera: Chrysopidae) for augmentative release against brown citrus aphid, *Toxoptera citricida* (Homoptera: Aphididae) in citrus. J Appl Entomol 125:383–388

Michaud JP, Alvarez R (2000) First collection of brown citrus aphid in Quintana Roo, Mexico. Fla Entomol 83:357

Michaud JP, Belliure B (2000) Consequences of foundress aggregation in *Toxoptera citricida* (Homoptera: Aphididae). Ecol Entomol 25:307–314

Michaud JP, Browning HW (1999) Seasonal abundance of the brown citrus aphid, *Toxoptera citricida* (Homoptera: Aphididae) and its natural enemies in Puerto Rico. Fla Entomol 82:424–447

Miller KA (1991) Response of Florida citrus growers to the freezes of the 1980s. Climate Res 1:133–144

Miller KA, Glantz MH (1988) Climate and economic competitiveness: florida freezes and the global citrus processing industry. Climatic Change 12:135–164

Moss GJ (1969) Influence of temperature and photoperiod on flower induction and inflorescence development in sweet orange (*Citrus sinensis* L. Osbeck). J Hortic Sci 44:311–320

Newman JA (2006) Using the output from global circulation models to predict changes in the distribution and abundance of cereal aphids in Canada: a mechanisitic modelling approach. Glob Chang Biol 12:1634–1642

Nickel O, Klingauf F (1985) Biologie und Massenwechsel der tropischen Citrus-Blattlaus *Toxoptera citricidus* in Beziehung zur Nützlingsaktivität und Klima in Misiones Argentinien (Homoptera: Aphididae). Entomol Gener 10:231–240 (in German, English sum.)

Ono S, Hirose K, Takahara T, Iwagaki I, Yoshinaga K (1988) Studies on physiological fruit drop in mid and late maturing citrus cultivars: 1. Inter-cultivar difference and relationship between physiological fruit drop and ecological factors. Bull Fruit Tree Res Stn D Kuchinotsu 10:47–68

Percy KE, Awmack CS, Lindroth RL et al. (2002) Altered performance of forest pests under atmospheres enriched by CO_2 and O_3. Nature 420:403–407

Qureshi JA, Rogers ME, Hall DG, Stansly PA (2009) Incidence of invasive *Diaphorina citri* (Hemiptera: Psyllidae) and its introduced parasitoid *Tamarixia radiate* (Hymenoptera: Eulophidae) in Florida citrus. J Econ Entomol 102:247–256

Qureshi JA, Stansly PA (2007) Integrated approaches for managing the Asian citrus psyllid *Diaphorina citri* (Homoptera: Psyllidae) in Florida. Proc Fla State Hort Soc 120:110–115

Qureshi JA, Stansly PA (2008) Rate, placement, and timing of aldicarb applications to control Asian citrus psyllid, *Diaphorina citri* (Hemiptera: Psyllidae) in oranges. Pest Manage Sci 64:1159–1169

Qureshi JA, Stansly PA (2009) Exclusion techniques reveal significant biotic mortalitysuffered by Asian citrus psyllid *Diaphorina citri* (Hemiptera: Psyllidae) populations in Florida citrus. Biol Control 50:129–136

Qureshi JA, Stansly PA (2010) Dormant season foliar sprays of broad-spectrum insecticides: an effective component of integrated management for *Diaphorina citri* (Hemiptera: Psyllidae) in citrus orchards. doi: 10.1016/j.cropro.2010.04.013

Reuther W (1973) Climate and citrus behavior. In: Reuther W (ed) The citrus industry, vol III. University of California, Division *of* Agricultural Sciences, Berkeley, CA, pp 280–337

Rocha-Pena MA, Lee RF, Lastra R, Niblett CL, Ochoa-Corona FM, Garnsey SM, Yokomi RK (1995) Citrus tristeza virus and its aphid vector, *Toxoptera citricida*. Threats to citrus production in the Caribbean and Central and North America. Plant Dis 79:437–445

Rosenzweig C, Phillips J, Goldberg R, Carrollh J, Hodges T (1996) Potential impacts of climate change on citrus and potato production in the US. Agric Syst 52:455–479

Seif AA, Islam AS (1988) Population densities and spatial distribution patterns of *Toxoptera citricida* (Kirk.) (Aphididae) in Citrus at Kenya coast. Insect Sci Appl 9:535–538

Stilling P, Rossi AM, Hungate B et al (1999) Decreased leaf-miner abundance in elevated CO_2: reduced leaf quality and increased parasitoid attack. Ecol Appl 9:240–244

Stone GN, Willmer PG (1989) Warm-up rates and body temperatures in bees: the importance of body size, thermal regime and phylogeny. J Exp Biol 147:303–328

Storey JM, Storey KB (1986) Winter survival of gall fly larva, *Eurosta solidaginis*: profiles of fuel reserves and cryoprotectants in a natural population. J Insect Physiol 32:549–556

Takanashi M (1989) The reproductive ability of apterous and alate viviparous morphs of the citrus brown aphid, *Toxoptera citricidus* (Kirkaldy) (Homoptera: Aphididae). Jpn J Appl Entomol Zool 33:266–269 (in Japanese, English summary)

Tang YQ, Lapointe SL, Brown LG, Hunter WB (1999) Effects of host plant and temperature on the biology of *Toxoptera citricida* (Homoptera: Aphididae). Environ Entomol 28:895–900

Tao CC, Tan MF (1961) Identification, seasonal population and chemical control of citrus aphids of Taiwan. J Agric Res 10:41–52

Tsai JH, Lee RF (1999) Biology and control of brown citrus aphid (*Toxoptera citricida* Kirkaldy) and citrus tristeza. Radcliffs IPM world text book. University of Minnesota, MN, USA

Tsai JH, Liu YH, Wang JJ, Lee RF (2000) Recovery of orange stem pitting strains of citrus tristeza virus (CTV) following single aphid transmission with *Toxoptera citricida* from a Florida decline isolate of CTV. Proc Fla State Hort Soc 113:75–78

Tsai JH, Wang KH (1999) Life table study of brown citrus aphid (Homoptera: Aphididae) at different temperatures. Environ Entomol 28:412–419

Turrell FM (1972) The science and technology of frost protection. In: Reuthcr W (ed) The citrus industry, vol. 3. University of Galifornia Press, Berkeley, CA, pp 338–446

Weathersbee AA III, Mckenzie CL, Tang YQ (2004) Host plant and temperature effects on *Lysiphlebus testaceipes* (Hymenoptera: Aphidiidae), a native parasitoid of the exotic brown citrus aphid (Homoptera: Aphididae). Ann Entomol Soc Am 97:476–480

Wiltbank WJ, Oswalt TW (1987) Low temperature killing points of citrus leaves during the 1984–85, 1985–86 and 1986–87 low temperature periods in Florida. Proc Fla State Hort Soc 100:113–115

Winterling G (1984) Freeze of 83 ... and the effects: a major shift in Florida's citrus belt. Weatherwise 37:305–306

Yamamura K, Kiritani K (1998) A simple method to estimate the potential increase in the number of generations under global warming in temperate zones. Appl Entomol Zool 33:289–298
Yin J, Schlesinger ME, Stouffer RJ (2009) Model projections of rapid sea-level rise on the northeast coast of the United States. Nat Geosci 2:262–266
Yocum GD, Joplin KH, Denlinger DL (1991) Expression of heat shock proteins in response to high and low temperatures extremes in diapausing pharate larvae of the gypsy moth, *Lymantria dispar*. Arch Insect Biochem Physiol 18:239–249

Chapter 7
Ecological Factors Influencing Pea Aphid Outbreaks in the US Pacific Northwest

Stephen L. Clement, Damon S. Husebye, and Sanford D. Eigenbrode

Abstract This chapter documents the history of pea aphid (*Acyrthosiphon pisum* (Harris) (Hemiptera: Aphididae)) outbreaks in the U.S. Pacific Northwest, with particular attention to the periodicity of outbreaks on field peas (*Pisum sativum* L.) in the Palouse region of eastern Washington over a 26 year period (1983–2008). In the Palouse region, grain legume crops are devastated by pea aphid feeding damage and legume viruses during outbreak years. Various abiotic and biotic factors and their possible controlling influence on changes in pea aphid densities, with emphasis on winter temperatures within the context of climate change, are presented and discussed. Mild winters, long thought to herald spring pea aphid outbreaks, are defined and related to outbreak cycles. While the 26-year database does not demonstrate a consistent link between mild winters and pea aphid outbreaks, it reveals a certain periodicity with outbreaks occurring every 6–9 years in the Palouse region. Warming winter temperatures over several years bridging the 20th and 21st centuries could potentially compress the frequency of outbreaks by accelerating aphid population development on legumes that provide migrants that colonize peas and other spring-sown food legumes in this region. Large plantings of alfalfa (*Medicago sativa* L.) 60–120 km west and southwest of the Palouse region are thought to be the source of alate aphids that migrate northeast and attain outbreak densities in some years. In conclusion, this chapter delivers a new long-term data set that will improve our understanding of how different abiotic and biotic factors influence the life-history processes of pest aphids.

Keywords *Acyrthosiphon pisum* · Outbreak cycles · Grain legumes · Climate change · US Pacific Northwest

S.L. Clement (✉)
USDA-ARS, Plant Germplasm Introduction and Testing Research Unit, Washington State University, Pullman, WA 99164-6402, USA
e-mail: slclement@wsu.edu

P. Kindlmann et al. (eds.), *Aphid Biodiversity under Environmental Change*,
DOI 10.1007/978-90-481-8601-3_7, © US Government 2010

7.1 Introduction

Insects as poikilotherms are dependent on temperature for life-history processes with their survival potentially impaired by extremes of low and high temperatures (Skirvin et al. 1997; Harrington et al. 2001; Harrington 2002; Klowden 2002). Thus, the prospect of global climate change begs the question of how insects will respond to possible changes in ambient temperatures. This question has been addressed by several investigators in recent years, leading to wide consensus that rising temperatures could affect insect development, survival, abundance, movement patterns, and insect–natural enemy interactions. Indeed, global mean surface temperatures have increased by ~0.6°C in the last century (Houghton et al. 2001). Therefore, changes in insect life-history processes in response to temperature changes could lead to increases in pest abundance and, by extension, more crop losses (Lawton 1995; Harrington et al. 2001; Bale et al. 2002; Harrington 2002; Cocu et al. 2004). Insects currently limited by high temperatures are not expected to respond to global warming or to experience population decreases (Harrington 2002).

For aphid pests of temperate crops, which tend to be limited more by low than by high temperatures, warming temperatures are expected to result in higher reproductive rates because of their short generation times and low developmental threshold temperatures. Moreover, prolonged favorable temperatures are conducive for advances in aphid phenology and rapid increases in aphid densities (Wellings and Dixon 1987; Harrington 2002; Cocu et al. 2004). Therefore, aphids, as sensitive indicators of temperature change, are ideal subjects for studying the impacts of climate change on insects (Harrington 2002; Harrington et al. 2007). In this regard, the pea aphid, *Acyrthosiphon pisum* (Harris) (Hemiptera: Aphididae), a major pest of grain legumes in the U.S. Pacific Northwest (Young et al. 1994; Clement et al. 2000), is an excellent candidate to study because ambient temperature is widely thought to be the most significant exogenous factor influencing the earliness of population increases that lead to spring and early summer outbreaks on grain legumes in this region (Eichmann and Webster 1940; Rockwood and Reeher 1943; Cooke 1963; Hampton 1983; Homan et al. 1992). Given the sensitivity of *A. pisum* to temperature change, the prospect of warmer winter temperatures could lead to more outbreaks and increased aphid-induced yield losses on spring-sown grain legumes in the Pacific Northwest.

The pea aphid exists in different strains and host races of genetic origin (Müller 1980; Via 1991; Sandström 1994; Bommarco and Ekbom 1996; Bournoville et al. 2000; 2004; Ferrari et al. 2006; Peccoud et al. 2009). For example, both red and green strains exist with the red form usually found on red clover and alfalfa and the green on alfalfa (*Medicago sativa* L.), pea (*Pisum sativum* L.), and red clover (*Trifolium pratense* L.) (Markkula 1963). We have found only the green strain (subspecies *destructor*) on alfalfa and peas in the Blue Mountain and Palouse regions of the U.S. Pacific Northwest (Fig. 7.1), although a "pink form" of *A. pisum* has been found in alfalfa fields in other U.S. regions (Hodgson 2007; Rethwisch 2008).

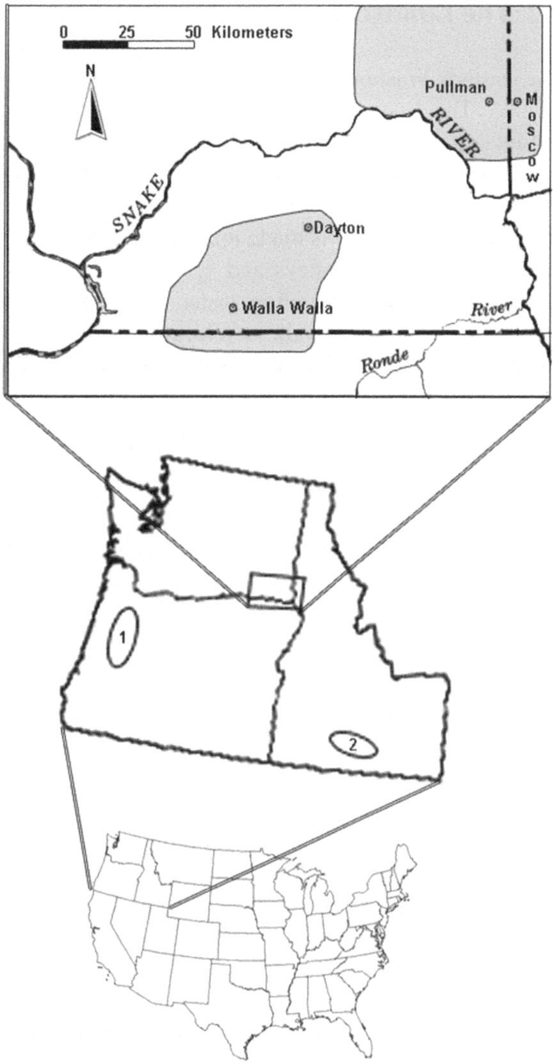

Fig. 7.1 United States (*lower panel*) with enlargements of U.S. Pacific Northwest (Oregon, Washington, Idaho) (*middle panel*) and the Blue Mountain/Columbia Basin region of eastern Oregon and Washington (Walla Walla) and Palouse region of eastern Washington (Pullman) (*upper panel*). Locations of pea aphid outbreaks in western Oregon (no. 1, *middle panel*) and southern Idaho (no. 2, *middle panel*)

In this chapter we document the history of pea aphid outbreaks in the U.S. Pacific Northwest, with specific attention to the periodicity of outbreaks on field peas in the Palouse region of eastern Washington (Fig. 7.1) over 26 years (1983–2008). This historical account is discussed in relation to pea aphid–host plant relationships in eastern Washington and the ecological factors that could influence irruptive changes in pea aphid densities. We offer suggestions on how specific factors could influence pea aphid dynamics, with particular attention to winter temperature patterns over 26 years and their possible influence on the periodicity of outbreaks in the U.S. Pacific Northwest.

7.2 The Landscape

The eastern Washington areas of interest are topographically and agriculturally diverse. The first area encompasses a major crop production area of the Blue Mountain/Columbia Basin region (~1,970 km^2), which extends from the south bank of the Snake River to Walla Walla, Washington (46°3'N, 118°21'W), a city in the heart of this region, and into central Oregon in a south/southwesterly direction (Fig. 7.1). This region is made up of several mountain ranges that are separated by geologically faulted valleys and synclinal basins, with rivers draining the mountainous slopes and providing water for irrigated crops (alfalfa, grain legumes, grapes, and onions) in valleys (elevations 282–1,067 m). Peas in this region are cultivated mainly as green peas for the processing and canning industry, with hectarage declining from 6,500 in 1990 to below 3,000 in 2002 and subsequent years (USDA, Washington Agricultural Statistics Service, Olympia, Washington; http://www.nass.usda.gov/wa).

The Palouse region, 60–100 km northeast of Walla Walla and covering 2,400 km^2 (elevations to 900 m), extends into northern Idaho (Fig. 7.1) and contains some of the most fertile and productive topsoil in the world for non-irrigated production of wheat, barley, lentils, and peas for dry seed (field peas) production (Muehlbauer et al. 1983; 1995; Black et al. 1998). Most of the field peas in the Palouse are grown within 50 km of Pullman, Washington and Moscow, Idaho (46°73'N, 117°9'W). Field pea hectarage is declining in the Palouse, going from yearly averages of ~48,000 hectares in the early 1980s to ~32,000 hectares per year in more recent years (2002–2007) (USDA, Washington Agricultural Statistics Service, Olympia, Washington; http://www.nass.usda.gov/wa).

In both regions, the climate is considered temperate with hot, dry summers following relatively wet winters and damp springs. Most precipitation falls as rain though substantial snow can fall in some winters in the mountains of the Blue Mountain region and on the Palouse (Black et al. 1998; NOAA: Western Regional Climate Center, http://www.wrcc.dri.edu/narratives/Washington.htm).

7.3 Pea Aphid – Legume – Relationships

Although active forms (nymphs, adults) of the pea aphid can possibly overwinter in France (Bournoville 1973) and survive mild British winters (McVean et al. 1999), this aphid mainly overwinters as diapausing eggs in perennial legumes in northern areas of the world (Bronson 1935; Dunn and Wright 1955; Markkula 1963; Sandström 1994; Bommarco and Ekbom 1995). In the U.S. Pacific Northwest, the aphid overwinters in the egg stage on alfalfa and other perennial legumes, or as active forms on perennial legumes in warmer areas or protected places (Eichmann 1940; McWhorter and Cook 1958; Cooke 1963; Johansen et al. 1979; Homan et al. 1992). For example, Eichmann and Webster (1940) found pea aphids "in small numbers in alfalfa at Pullman, Washington, during each month of the mild winter of 1937–1938." In the Blue Mountain region, April–May migratory flights from alfalfa provide the aphids that colonize spring peas in this region (Eichmann and Webster 1940; McWhorter and Cook 1958; Cooke 1963; Clement and Eigenbrode, unpublished). Moreover, pea aphid outbreaks on these Blue Mountain spring peas follow high survival of overwintering eggs and active forms on alfalfa (Cooke 1963).

The role of alfalfa in providing pea aphids that colonize peas in southern Idaho (Fig. 7.1) is well-established (Hampton 1983; Hampton and Weber 1983a, b). Likewise, this pattern of overwintering eggs in alfalfa and other perennial legumes giving rise to first-generation individuals that produce alate migrants and colonizers of spring legumes, such as peas, is grounded in the European literature (Sandström 1994; Bommarco and Ekbom 1995, 1996; McVean et al. 1999). In Sweden, for example, all pea aphids on peas are green and develop from alate migrants of the green form that originate in alfalfa (Sandström 1994, 1996; Bommarco and Ekbom 1996).

By contrast, the overwintering hosts for the pea aphids that appear in the spring and early summer on field peas in the Palouse have not been identified, although alfalfa and other perennial legumes have been listed as principal sources of these aphids (Halfhill and Featherston 1973; Hampton and Weber 1983a; Homan et al. 1992). If we accept the notion that large numbers of pea aphids depart alfalfa and other perennial legumes for field peas on the Palouse, we need to identify the geographical locations of these legumes. Perhaps alfalfa fields in the Columbia Basin/Blue Mountain region and other areas 60–120 km west and southwest of Pullman (Fig. 7.1) are the source of alate aphids that migrate northeast and attain outbreak densities in some years on Palouse field peas. This is a reasonable assumption because these areas to the west and southwest of the Palouse support large plantings of alfalfa, with yearly averages approaching 39,000 hectares in Walla Walla and adjoining regions compared to yearly averages of 4,900 hectares for the Palouse region around Pullman (estimates for 1985–2007) (USDA, Washington Agricultural Statistics Service, Olympia, Washington; http://www.nass.usda.gov/wa). In addition, warmer winters in the Blue Mountain region (Table 7.1) favor high survival rates for overwintering populations of the pea aphid. Because the pea aphid can move long distances when assisted by winds (Eichmann and Webster 1940; Halfhill and Featherston 1973; Johansen et al. 1979), it is plausible that prevailing spring and early summer winds, which generally come from the west-southwest (National Climatic Data Center, National Oceanic and Atmospheric Administration, Asheville, North Carolina; http://www.ncdc.noaa.gov), assist with the migration of pea aphids from the Blue Mountain region to the Palouse region. Red clover, a chief host of pea aphid in eastern North America (Via 1991), is an unlikely source of migrant pea aphids in eastern Washington because it occurs as scattered plants in fields and along roads in areas between the Blue Mountain/Columbia Basin region and the Palouse (Clement and Eigenbrode, unpublished surveys).

Additional evidence to link alfalfa with subsequent pea aphid infestations on Palouse peas comes from the occurrence of three pathogenic, aphid-transmitted viruses in alfalfa. Two of these viruses are transmitted non-persistently (*Alfalfa mosaic virus*, *Pea streak virus*) and one persistently (*Bean leaf roll virus*) (Hampton and Weber 1983a, 1983b; Hagedorn 1984; Homan et al. 1992; Kaiser et al. 2000;

Table 7.1 December–February temperature patterns in the Blue Mountain (Walla Walla) and Palouse (Pullman) regions of eastern Washington and their relationship to pea aphid outbreak and non-outbreak years over a 26 year period in the Palouse region[a]

Year	No. of days < −10°C[b]		Mean low temperatures (°C)[b]		Outbreak (+), No-outbreak (−)	
	Walla Walla	Pullman	Walla Walla	Pullman	Expected[c]	Actual
1982–1983	0	7	−0.8	−1.6	±	+
1983–1984	16	19	−3.7	−4.7	−	−
1984–1985	17	29	−5.8	−7.9	−	−
1985–1986	13	26	−4.6	−4.9	−	−
1986–1987	5	13	−2.4	−3.4	−	−
1987–1988	7	13	−3.1	−3.9	−	−
1988–1989	10	19	−4.4	−5.7	−	−
1989–1990	2	6	−1.1	−2.4	−	+
1990–1991	13	14	−4.5	−4.1	−	−
1991–1992	0	0	0.1	−0.4	+	−
1992–1993	24	23	−6.3	−6.2	−	−
1993–1994	2	4	−1.1	−2.1	−	−
1994–1995	3	12	−1.1	−2.2	−	−
1995–1996	9	17	−3.7	−4.4	−	+
1996–1997	8	13	−2.7	−3.8	−	−
1997–1998	3	5	−1.5	−2.4	−	−
1998–1999	6	9	−0.7	−3.0	−	−
1999–2000	0	0	0.0	−2.5	+	−
2000–2001	0	9	−2.2	−5.1	−	−
2001–2002	0	7	−1.8	−3.7	−	−
2002–2003	0	2	0.1	−1.2	+	−
2003–2004	6	7	−2.7	−3.2	−	−

Table 7.1 (continued)

Year	No. of days < –10°C[b]		Mean low temperatures (°C)[b]		Outbreak (+), No-outbreak (–)	
	Walla Walla	Pullman	Walla Walla	Pullman	Expected[c]	Actual
2004–2005	10	6	−3.6	−3.1	–	+
2005–2006	10	16	−2.8	−3.2	–	–
2006–2007	8	8	−3.4	−3.9	–	–
2007–2008	6	8	−2.4	−4.2	–	–

[a]Temperature data from Whitman Mission, U.S. National Park Service, Walla Walla, Washington, and the National Climatic Data Center, National Oceanic and Atmospheric Administration, Asheville, North Carolina (http://www.ncdc.noaa.gov)
[b]No days below –10°C and mean low temperatures greater than 0°C portend a mild winter
[c]Expected outbreak in the Palouse is based on criteria (Walla Walla records) in preceding footnote

Clement 2006). In 2006, over 70% of alfalfa fields sampled in the Blue Mountain region tested positive for *Alfalfa mosaic virus* and *Bean leaf roll virus* (Eigenbrode, Papu, Clement, unpublished data). However, alfalfa is not a host for *Pea enation mosaic virus* (Larsen et al. 1996; Eigenbrode, unpublished), another persistently transmitted virus and one of the most important pathogenic diseases (with *Bean leaf roll virus*) of peas and other grain legumes during virus epidemic years in the Palouse region (Klein et al. 1991; Clement 2006). Thus, before pea aphids can transmit *Pea enation mosaic virus* to peas they must acquire it from reservoir hosts that, to date, have not been identified (Larsen et al. 1996). Austrian winter pea, *P. sativum* L. ssp. *arvense*, was listed as a likely winter reservoir in the Blue Mountain region in the early 1960s (Cooke 1963), with wild perennial legumes another possibility.

What is the fate of pea aphids on Palouse peas as the crop reaches maturity and dries in the summer? Do winged forms migrate to nearby alfalfa fields or do they die in the crop or soon after leaving it? Cooke (1963) addressed these questions while studying the pea aphid in the Blue Mountain region from 1947–1960, and Eichmann and Webster (1940) observed return migrations of pea aphids to alfalfa in summers in the same region. The cropped landscape of the Palouse dries out rapidly from late July to mid-August, when aphids appear to perish on the drying legume plants, perhaps with the help of natural enemies (Clement and Eigenbrode, personal observations). We have considered the possibility that alates disperse from the annual legumes during this dry down period and colonize the widely dispersed fields of alfalfa in the Palouse region; however, our late fall and early spring surveys of alfalfa fields and winter legumes in the Palouse have not detected pea aphids. Thus, the Palouse pea crop appears to serve principally as a "sink" for aphids, in contrast to the situation in Europe where peas are considered a "source" host that pea aphids readily exploit before moving to other habitats and leguminous hosts as the crop matures (Bommarco and Ekbom 1996; McVean et al. 1999). A return migration to alfalfa and other leguminous hosts in the Blue Mountain region would have to be accomplished in the face of prevailing summer winds that blow mostly from a westerly direction (Cooke 1963), an unrealistic scenario in our opinion.

7.4 Characterizing Outbreaks

Rockwood and Reeher (1943) defined spring outbreaks of the pea aphid in western Oregon (Fig. 7.1) as "the occurrence of countless millions of aphids and observable aphid damage in many fields over a wide area." Other entomologists in the Pacific Northwest have associated the term "outbreak" with the appearance of destructive and high temporal densities of pea aphids (Eichmann and Webster 1940; Cooke 1963), or they used the term without defining it (Eichmann 1940). We define a pea aphid outbreak, after Berryman's (1987) general definition, as the relatively sudden appearance of very high densities of pea aphids on field peas over a wide area.

Using pea aphid counts on plants in commercial pea fields, published accounts, and personal communications from scientists, Clement (2006) devised a numerical index that revealed 6–9 year intervals (1983–2005) between outbreak years on the

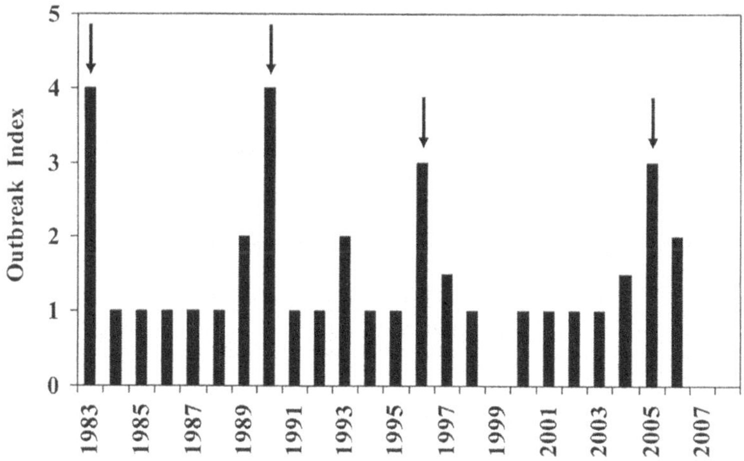

Fig. 7.2 Twenty-six year history of pea aphid outbreaks on field peas in the Palouse region of eastern Washington. Index values of 3–4 reflect outbreak years (*arrows*). Fields were devoid of pea aphids in 1999, 2007, and 2008

Palouse. Figure 7.2, revised and expanded from Clement (2006), depicts the history of outbreaks in the Palouse between 1983 and 2008 (no aphids detected in 1999, 2007, 2008). For this index, a value of 0 approaching 1 = aphids absent or "few" counted on plants; 1 approaching 2 = high counts (peak averages of 10–99 aphids per plant) in 26–50% of the sampled fields; 2 approaching 3 = very high counts (peak averages >100 aphids per plant) in 51–75% of the fields; and 3 approaching 4 = peak counts averaging >100 aphids per plant in 76–100% of the fields (Clement 2006). The very high aphid densities that characterize index values 3–4 represent outbreaks in this analysis. These outbreak densities far exceed recommended insecticide treatment thresholds for pea aphids on peas (see Homan et al. 1992).

A "quick in situ aphid count method" was used to generate the pea aphid numerical indices in Fig. 7.2 (Clement 2006). The method consisted of visually counting all aphids on 10–30 plants (1–2 min per plant) at 1–2 m intervals along well-spaced transects (1–4 per field). Sampling was done on 3–25 unsprayed fields per year at various time intervals during the major developmental period (mid May–early July) of the commercial pea crop in the Palouse region. The sampling intensity was reduced in years in which few or no aphids could be found in the crop. All commercial fields were in Whitman County, Washington, and within 30 km of Pullman.

The pea aphid count method was sufficient to identify the onset and duration of pea aphid infestations and outbreak densities. Illustrative of the sudden onset of outbreak densities are the pea aphid counts from 1990 and 2005, which increased from 6.5 to 14 aphids per plant in early June to over 100 aphids per plant in mid-June of both years in Palouse pea fields (Fig. 7.3). Figure 7.3 also shows the aphid density patterns for three non-outbreak years (1989, 1995, 2006), characterized by peak densities below 99 aphids per plant (1–2 index values). Pea aphid outbreaks

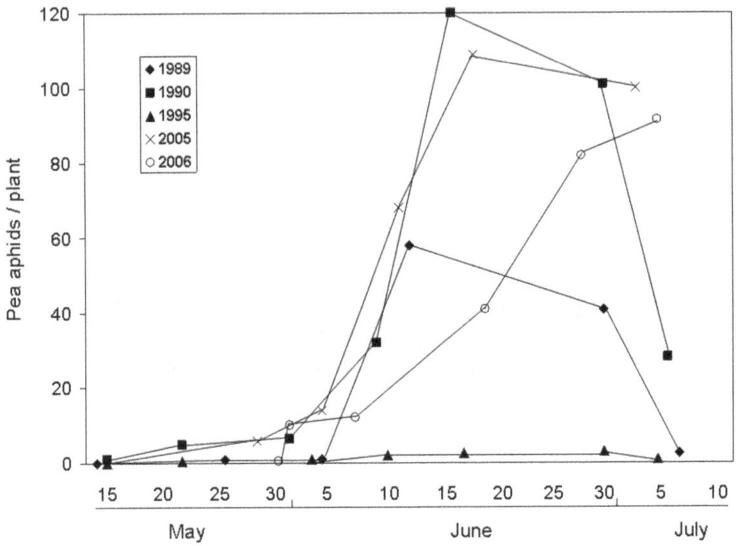

Fig. 7.3 Changes in pea aphid densities (aphids/plant; 30–180 plants/data point) in pea fields during two outbreak (1990, 2005) and three non-outbreak years (1989, 1995, 2006), Palouse region of eastern Washington

and virus epidemics are not always linked as revealed by the absence of outbreak densities in 1998 (index of 1; Fig. 7.2) when Palouse pea fields exhibited widespread symptoms of aphid-vectored viruses (S. Eigenbrode, unpublished data).

7.5 History of Outbreaks

Pea aphid outbreaks in the Pacific Northwest were first reported by Rockwood and Reeher (1943), who documented spring outbreaks in 1917–1918 and eight more between 1919 and 1942 on fall sown annual legumes (Austrian winter peas, common vetch, *Vicia sativa* L.) in western Oregon (Fig. 7.1). In the Blue Mountain region of eastern Washington and eastern Oregon (Fig. 7.1), aphid outbreaks were first noted in 1934 on spring (green) peas cultivated for the canning industry. This infant industry, started in spring 1933, was almost destroyed by this outbreak (Eichmann and Webster 1940; Cooke 1963). In the Palouse region (Fig. 7.1), high year-to-year fluctuations in pea aphid densities occurred on field peas between 1983 and 2008, with devastating effects on peas and other grain legume crops during the outbreak years of 1983, 1990, 1996, and 2005 (Hagedorn 1984; Klein et al. 1991; Clement et al. 2000; Clement 2006) – see Fig. 7.2. This pattern is consistent with the observation that aphid outbreaks "rarely occur over successive years" (Wellings and Dixon 1987). With the exception of 3 years (1999, 2007, 2008), when commercial pea fields were devoid of pea aphids, there were always some Palouse fields supporting aphid populations in non-outbreak years (reflected by index values of 1–2 for 19 years; Fig. 7.2). Additionally, pea aphid outbreaks and virus epidemics have periodically occurred on peas in southern Idaho (Fig. 7.1) (Hampton 1983; Hampton and Weber 1983a, 1983b; Stoltz and Forster 1984).

7.6 Winter Temperatures and Outbreaks

Entomologists have proposed that minimum winter temperatures influence early pea aphid population increases that lead to outbreaks in the U.S. Pacific Northwest, with mild winters setting the stage for outbreaks via high survival of over-wintering nymphs and adults on legumes (Eichmann and Webster 1940; Rockwood and Reeher 1943; Cooke 1963; Hampton 1983; Homan et al. 1992). Rockwood and Reeher (1943), using temperature records and field notes for 29 years in western Oregon, first established an association between a "mild winter" in the Pacific Northwest and subsequent pea aphid outbreaks in this region. Later, Cooke (1963) reinforced that association, reporting that "severe outbreaks in the Blue Mountain area rarely, if ever, followed winter seasons in which subzero temperatures occurred between January 15 and February 10." Many present-day pea producers and crop consultants continue to associate mild winters with subsequent pea aphid problems on the Palouse (personal communications to the authors).

To examine whether mild winters herald outbreaks in the Palouse region, winter temperature records (1982–2008) from the Blue Mountain (Walla Walla) and Palouse (Pullman) regions were compared with pea aphid numerical indices. Records from both regions were used because we accept the possibility that aphid colonizers of Palouse field peas could originate locally and/or from the Blue Mountain region. Based on December–February temperature records, a "mild winter" was defined as one in which mean monthly low temperatures were $\geq 0°C$ and there were was an absence of days with temperatures $\leq -10°C$ (slightly modified from Rockwood and Reeher (1943)). Using these criteria, we found no consistent link between "mild winters" and ensuing aphid outbreaks. Based on Walla Walla temperatures, expected and actual outcomes (outbreak, no-outbreak) were incongruent or nearly so for seven of the 26 years (Table 7.1). Expected and actual outcomes were in agreement for 19 years or 73% of the observations. The mild winters of 1991–1992, 1999–2000, and 2002–2003 did not produce aphid outbreaks, however, aphid outbreaks followed the more severe winters of 1989–1990, 1995–1996, and 2004–2005 (Fig. 7.2, Table 7.1). For the winter of 1982–1983, the monthly temperature value was slightly below 0°C, but there were no days below −10°C (Table 7.1); therefore, categorizing this winter as "mild" in advance of the 1983 outbreak would raise the agreement rate between expected and actual outcomes to 77%. Similarly, Hampton (1983) was unable to relate winter temperatures to high pea aphid densities and transmission of Bean leaf roll virus to peas in southern Idaho. Compared to Walla Walla winters, Pullman winters were colder with only one "mild winter" (1991–1992) between 1983 and 2008, and it preceded an outbreak (Table 7.1). In conclusion, mild winters do not consistently precede outbreaks in the Palouse region, although there was good agreement between observed and predicted aphid outbreak events over a 26-year period.

If ambient temperatures continue to rise, as they have over the past 40 years in the Northern Hemisphere, including Western North America (Jones 1988; Rasmussen and Conway 2004; Robeson 2004), this pattern could plausibly lead to more pea aphid outbreak years in the U.S. Pacific Northwest. Indeed, there is an expectation by some aphidologists that milder winters, as one manifestation of global climate change, will result in higher densities of some species of aphids (Harrington 2002, 2003; Cocu et al. 2004) (but see Dixon 1998). However, rising temperatures in the U.S. Pacific Northwest may not portend more pea aphid outbreaks in the region, as evidenced by the warming trend in winter temperatures (Walla Walla and Pullman) over several years bridging the 20th and 21st centuries. During this period, the intervals between outbreaks have been relatively stable (Fig. 7.2). Yet, a warming trend is apparent upon comparison of two periods: 1982–1994 with averages of 9.1 (Walla Walla) and 14.4 (Pullman) winter days with temperatures below –10°C (two outbreaks occurred), compared to reduced averages of 4.9 (Walla Walla) and 8.5 (Pullman) winter days with temperatures below –10°C for the ensuing 14-years (1995–2008) when two outbreaks occurred (Table 7.1). Also, this warming trend is reflected by Walla Walla and Pullman winter temperatures, which rose ~0.9°C from 1982–1994 to the next 14-year period in both regions (Table 7.1).

Nonetheless, it is premature to rule out temperature as an important determinant of pea aphid outbreak cycles and to discount the impact of regional warming trends on pea aphid life-history events. Temperature can potentially interact with other ecological factors, both abiotic and biotic, to influence pea aphid outbreaks (see below) and the growth patterns and densities of other aphid species (Adler et al. 2007).

7.7 Other Causal Factors

Besides winter temperatures, other weather-related factors, natural enemy populations, farming practices, and plant quality attributes can potentially influence pea aphid life-history events and thus contribute to the occurrence of outbreaks in the Pacific Northwest (Table 7.2). Karley et al. (2004) previously reviewed the effect of three (weather conditions, natural enemies, and plant quality) of these factors on mid-season crashes of aphid populations on grasses and forbs.

Although winter temperatures alone do not consistently predict subsequent pea aphid outbreaks in the Palouse region, there is evidence that pea aphid winter mortality rates on alfalfa in the Pacific Northwest wax and wane with mild and cold winters (Rockwood and Reeher 1943; Cooke 1963). While cold winters increase aphid mortality and delay aphid population development, they also can delay planting of a pea crop that will lead to higher pea aphid densities over a long growing season, as McVean et al. (1999) observed in the U.K. In a similar vein, late-planted peas in the Blue Mountain region supported higher densities of pea aphid (Eichmann and Webster 1940). This is an example of how two general factors, weather and farming practice (late-planting), jointly effect pea aphid population processes.

Table 7.2 Ecological factors (documented and postulated) interacting to influence pea aphid population processes and outbreak events on spring-sown legumes in the U.S. Pacific Northwest

Factor			
General	Specific	Effect on aphid	References[a]
Weather	Winter temperatures	Influences winter mortality rates, density levels, activity cycles	Rockwood & Reeher (1943), Cooke (1963), Homan et al. (1992), McVean et al. (1999)
	Severe rainstorms	Lower densities	Eichmann & Webster (1940)
	Optimum/timely rainfall	Densities maintained/increased	Eichmann & Webster (1940), Rockwood & Reeher (1943)
	Wind speeds and trajectories	Timing/direction of migration	Eichmann & Webster (1940), Clement (2006)
	Severe windstorms	Lower densities	Eichmann & Webster (1940), McWhorter & Cook (1958)
Farming practice	Alfalfa cutting cycles	Influences mortality, alate production, emigration	Eichmann & Webster (1940), McWhorter & Cook (1958), Cooke (1963), Hampton (1983), Hampton & Weber (1983a)
	Delayed planting	Increases densities	Eichmann & Webster (1940), McVean et al. (1999)
	Crop area	Influences densities	Eichmann & Webster (1940), Dixon & Kindlmann (1990), McVean et al. (1999), Cocu et al. (2004)
Plant quality	Moisture and nutrients	Influences densities	Rockwood & Reeher (1943), Karley et al. (2004)
Natural enemies	Predators and parasitoids	Lower densities	Eichmann & Webster (1940), Cooke (1963), van den Bosch et al. (1966), Halfhill et al. (1972), Halfhill & Featherston (1973),

Table 7.2 (continued)

Factor			
General	Specific	Effect on aphid	References[a]
			Johansen & Eves (1973), Johansen et al. (1979), Frazer et al. (1981), Gutierrez et al. (1984), Ekbom (1994), Eigenbrode et al. (1998), White & Eigenbrode (2000), Rutledge et al. (2003), Snyder & Ives (2003), Chang & Eigenbrode (2004), Chang et al. (2004), Evans (2004)
	Pathogenic fungi	Lower densities	Rockwood (1950), Cooke (1963), Feng et al. (1990), Pickering & Gutierrez (1991), Bommarco & Ekbom (1996), White (1998), Duetting (2002)

[a]Most references are specific to *A. pisum*.

Rain and wind storms, common in the spring in the Pacific Northwest, can reduce pea aphid numbers by washing or knocking individuals from host plants (Eichmann and Webster 1940; McWhorter and Cook 1958; Eigenbrode, unpublished data). But timely and optimal rainfall in the fall and spring is important for supplying moisture to sustain suitable host plant growth for pea aphid population development during these seasons (Rockwood and Reeher 1943). Additionally, rainstorms and prevailing wind speeds and trajectories are thought to influence the timing and direction of spring migrations of winged pea aphids and subsequent outbreaks in eastern Washington (Eichmann and Webster 1940; Clement 2006). Thus, the amount and timing of rain and windstorms in early spring can influence, in various ways, pea aphid population dynamics on peas and other spring-planted annual legumes in the Pacific Northwest.

Farming practices other than planting date also impact pea aphid life-history processes. For example, alfalfa cutting can force winged aphids to take flight and colonize annual legumes (references in Table 7.2). If the alfalfa crop is of high quality and supports abundant aphid populations, the impact on annual legumes is likely greater. Thus, plant quality (succulent alfalfa plants) and a farming practice (alfalfa cutting) can become intertwined to influence pea aphid. Farmers also determine the amount of agricultural land they devote to specific crops, which in turn could influence aphid densities if high food resource abundance, on a landscape scale, is related to high pest aphid abundance (e.g., Dixon and Kindlmann 1990; Cocu et al. 2004). If this relationship holds, pea aphid densities would increase with increases in pea hectarage. However, this was not the case in U.K. where McVean et al. (1999) reported constant numbers of pea aphids per unit area of pea crop, despite fluctuating hectares planted to peas. Moreover, outbreak events and pea crop hectarage were unrelated over 26 years in the Palouse region, exemplified by the 2005 outbreak that coincided with only 31,000 hectares of peas and when three previous outbreaks occurred during years when 42,000–50,000 hectares were devoted to field pea production.

Many reports have implicated natural enemies as key biotic determinants of pea aphid density levels in Pacific Northwest alfalfa fields, with a wide variety of natural enemies (coccinellids, hemipteran predators, syrphid larvae, hymenopterous parasitoids, fungal diseases) found in these fields (Eichmann and Webster 1940; Cooke 1963; Halfhill et al. 1972; Halfhill and Featherston 1973; Johansen and Eves 1973; Johansen et al. 1979). However, there is no experimental evidence linking the action of these natural enemies with reductions in pea aphid numbers in alfalfa fields in the Pacific Northwest.

Can we look to results from aphid–natural enemy interaction research in other geographical areas for insight into potential impacts of natural enemies on density levels of pea aphid in both alfalfa and pea fields in the Pacific Northwest? Ekbom (1994) reported that polyphagous predators, including *Coccinella septempunctata* L. (Coleoptera: Coccinellidae), had no significant impact on pea aphid numbers in alfalfa, clover, and pea fields in Sweden. In perennial crops in California, coccinellids reduced pea aphid numbers but they did not prevent the build-up of high populations (Gutierrez et al. 1984). On the other hand, coccinellid predators appeared to have a continuous and marked effect on pea aphid numbers in Vancouver, Canada (Frazer et al. 1981), and the seven-spotted ladybird beetle, *Coccinella septempunctata* L. (Coleoptera: Coccinellidae), dampened the potential of pea aphid populations to reach high numbers in Utah alfalfa fields (Evans 2004). On the Palouse, coccinellid predators demonstrably reduced pea aphid populations on peas in field enclosures (Chang and Eigenbrode 2004), yet pea aphid populations reached high densities in the presence of coccinellid predators in other studies in this region (White and Eigenbrode 2000; Rutledge et al. 2003).

The suppressive effect of parasitoids on pea aphid densities has been investigated in the U.S., including the Pacific Northwest. Although van den Bosch et al. (1966) suggested that *Aphidius smithi* Sharma and Subba Rao (Hymenoptera:Aphidiidae) controlled this aphid in California, the biological control effectiveness of its congener, *A. ervi* Haliday, was not established after detailed field experiments in Wisconsin alfalfa fields (Snyder and Ives 2003). Although *A. ervi* is the predominant parasitoid of pea aphid in the Palouse region, its typically low attack rate (< 10% parasitism) appears to have little impact on aphid population levels (White and Eigenbrode 2000; Rutledge et al. 2003; Chang et al. 2004). Moreover, inundative releases of hymenopterous parasitoids were shown to be impractical for controlling pea aphid on alfalfa in the Pacific Northwest (Halfhill and Featherston 1973).

Entomopathogenic fungi, notably *Pandora neoaphidis* (Remaudière and Hennebert) Humber, might have potential to reduce pea aphid populations, as suggested by research in alfalfa in California (Pickering and Gutierrez 1991). However, *P. neoaphidis* is evidently enzootic in the Pacific Northwest (M-G. Feng, A. Hajek, L. Lacey, personal communications) where low infection rates occur (Rockwood 1950; White 1998; Duetting 2002; Eigenbrode, personal observations). Dry weather conditions appear to curtail epizootic events in this region (Rockwood 1950; Feng et al. 1990; Duetting 2002).

Our review of the pertinent literature on pea aphid–natural enemy interaction research (Table 7.2) indicates that natural enemies do not consistently regulate pea aphid populations in the Pacific Northwest. To be sure, the impact of natural enemies on pest aphids, in general, is uncertain (reviewed by Wellings and Dixon 1987). Furthermore, predators, parasitoids, and entomopathogens are unlikely to respond quickly enough to prevent pea aphid outbreaks on peas in the Palouse region where aphid numbers suddenly increase over 2 weeks in outbreak and some non-outbreak years (Fig. 7.3). This rapid population growth is not surprising given that pea aphids, under optimal temperatures, can develop from first instars to mature and reproducing adults in as little as 10 days (Hutchison and Hogg 1984). Coccinellid populations and percent parasitism of pea aphids in Palouse field peas tend to lag behind explosively growing aphid populations in these fields (White and Eigenbrode 2000; Chang et al. 2004), further exacerbating the potential for an outbreak.

The arrival of new coccinellid species in the Pacific Northwest in the 1990s, namely *C. septempunctata* and *Harmonia axyridis* (Pallas) (Elberson 1992; LaMana and Miller 1996; Eigenbrode et al. 1998; White and Eigenbrode 2000; Clement et al. 2004; Snyder et al. 2004), calls for a detailed re-assessment of pea aphid–coccinellid interactions in pea fields, and in alfalfa fields that may harbor year-round aphid populations. Can these introduced coccinellids, along with other natural enemies, exert density-dependent control of pea aphid numbers in alfalfa during periods leading up to the planting of spring legumes, and even in late-summer and fall as annual legume crops mature and dry down? Support for this possibility was provided by Ekbom (1994) who found that natural enemies were more effective in regulating pea aphid numbers on perennial legumes than on annual peas in Sweden.

7.8 Forecasting Outbreaks

While our 26-year database does not demonstrate a consistent link between mild winters and pea aphid outbreaks, it reveals a certain periodicity to the occurrence of outbreaks in the Palouse region (Fig. 7.1). More evidence that temperature plays a role in these events comes from the absence of outbreaks after colder winters in 19 out of 26 years (Table 7.1; Clement 2006). Thus, a temperature-driven analytical model might help us better predict the frequency and regularity of outbreaks, which would help farmers with timely and informed pea aphid control decisions.

Although weather data have been used in research and development of analytically-based systems to predict impending aphid outbreaks (Wellings and Dixon 1987; Dixon 1998; Cocu et al. 2004; Thackray et al. 2004), previous attempts to model pea aphid activity cycles using aphid counts and weather data were not successful in Sweden (Bommarco and Ekbom 1995) and Germany (Thacker et al. 1997). We take this as evidence that multiple biotic and abiotic inputs are required for optimal model development. To generate these inputs, more study of the ecology of pea aphids and more long-term data collection on the interactive abiotic and biotic factors affecting this aphid in legume crops (Table 7.2) in the Pacific Northwest is required.

Active monitoring of pea aphids in source legumes and their subsequent flight activity patterns might improve our ability to anticipate outbreaks. Although Bommarco and Ekbom (1995) found that suction trap catches were not useful for predicting pea aphid infestations on peas in Sweden, these traps could help track aphid movement and population behavior patterns during periods of changing ambient temperatures. Suction trap counts have provided a measure of pea aphid abundance on peas in the U.K. (McVean et al. 1999) and have yielded important information about the geographical patterns of other species of aphids and the association of these patterns with European climatic variables, with an eye towards understanding how aphids respond to global climate changes (Cocu et al. 2005a, 2005b; Harrington et al. 2007).

7.9 Concluding Remarks

An obvious but expected message of this chapter is that temperature is important to the population dynamics of the pea aphid in the U.S. Pacific Northwest and this climatic factor influences the periodicity of outbreaks of this pest aphid in this region. But equally this volume makes it clear that temperature, and mild winters in particular, is not the sole determinant of pea aphid outbreak events. Changes in the frequency and severity of pea aphid outbreaks in this region have not occurred in response to a moderation in winter temperatures over a number of years bridging the 20th and 21st centuries. Additionally, this chapter, while examining a complex system involving one temperate aphid species in a topographically and ecologically diverse part of the world, reinforces the fact that ecological factors (biotic and abiotic), linked to and interacting with temperature, influence the life-history processes of poikilothermic insects.

In the context of identifying and understanding how interacting abiotic and biotic factors influence pea aphid life-history processes, we can only generalize about their overall effect (including climate change) on this aphid in the U.S. Pacific Northwest without an expanded base of knowledge. To paraphrase Harrington et al. (2001) and Harrington (2003), entomologists and other scientists have had very little success generalizing predictions of climate change impacts on insects and, specifically, how weather-related factors influence aphid population dynamics. Moving beyond this current state will require more long-term data sets like the ones generated by the

Rothamsted Insect Survey (Woiwod and Harrington 1994; Harrington 2002) and researchers at the Lithuanian Institute of Agriculture (Smatas et al. 2008), which will make it possible to study aphid population trends and their relationships to weather variables and other factors. This chapter delivers a new set of pest aphid and weather data. These data can be a springboard for further inquiries into the causes of pea aphid outbreaks, in addition to improving our overall understanding of how single and multiple factors influence the life-history processes of pest aphids.

Acknowledgements We are grateful to Leslie Elberson for unwavering help and assistance with the preparation of this chapter. This work was partially supported by grants to S.D. Eigenbrode (USDA-NRI and USDA-CSREES Risk Assessment and Mitigation Program) (Award No. 2008-51101-0452).

References

Adler LS, De Valpine P, Harte J, Call J (2007) Effects of long-term experimental warming on aphid density in the field. J Kans Entomol Soc 80:156–168

Bale JS, Masters GJ, Hodkinson ID, Awmack C, Bezemer TM, Brown VK, Butterfield J, Buse A, Coulson JC, Farrar J, Good JEG, Harrington R, Hartley S, Jones TH, Lindroth RL, Press MC, Symrnioudis I, Watt AD, Whittaker JB (2002) Herbivory in global climate change: direct effects of rising temperatures on insect herbivores. Glob Chang Biol 8:1–16

Berryman AA (1987) The theory and classification of outbreaks. In: Barbosa P, Schultz JC (eds) Insect outbreaks. Academic Press, Inc., San Diego, pp 3–27

Black AE, Strand E, Morgan P, Scott JM, Wright RG, Watson C (1998). Biodiversity and land-use history of the Palouse bioregion: pre-European to present, Chapter 10. In: Sisk TD (ed) Perspectives on land-use history of North America: a context for understanding our changing environment. U.S. Geological Survey, Biological Resources Division, Biological Science Report USGS/BRD/BSR. http://biology.usgs.gov/luhna/ . Accessed 2 Aug 2006

Bommarco R, Ekbom B (1995) Phenology and prediction of pea aphid infestations on peas. Int J Pest Manage 41:109–113

Bommarco R, Ekbom B (1996) Variation in pea aphid population development in three different habitats. Ecol Entomol 21:235–240

Bournoville R (1973) Observations écologiques sur l'hivernation du puceron du pois *Acyrthosiphon pisum* (Harris) et de ses parasites dans la region de Versailles. Ann de Zool Ecol Anim 5:13–28

Bournoville R, Carre S, Badenhausser I, Simon JC, Hennis C, Greze C (2004) Host-races of the pea aphid, *Acyrthosiphon pisum*: biological criteria and feeding behaviour of clones originating from legumes. In: Simon JC, Dedryver CA, Rispe C, Hullé M (eds) Aphids in a new millennium. INRA Editions, Paris, pp 413–419

Bournoville R, Simon JC, Badenhausser I, Girousse C, Guilloux T, Andre S (2000) Clones of the pea aphid, *Acyrthosiphon pisum* (Hemiptera: Aphididae) distinguished using genetic markers, differ in their damaging efficiency on a resistant alfalfa cultivar. Bull Entomol Res 90:33–39

Bronson TE (1935) Observations on winter survival of pea aphid eggs. J Econ Entomol 28: 1030–1036

Chang GC, Eigenbrode SD (2004) Delineating the effects of a plant trait on interactions among associated insects. Oecologia 139:123–130

Chang GC, Neufeld J, Duetting PS, Eigenbrode SD (2004) Waxy bloom in peas influences the performance and behavior of *Aphidius ervi*, a parasitoid of the pea aphid. Entomol Exp Appl 110:257–265

Clement SL 2006. Pea aphid outbreaks and virus epidemics on peas in the U.S. Pacific Northwest: histories, mysteries, and challenges. Online. Plant Health Progress doi:10.1094/PHP-2006-1018-01-RV

Clement SL, Elberson LR, Youssef N, Young FL, Evans MA (2004) Cereal aphid and natural enemy populations in cereal production systems in eastern Washington. J Kans Entomol Soc 77:165–173

Clement SL, Wightman JA, Hardie DC, Bailey P, Baker G, McDonald G (2000) Opportunities for integrated management of insect pests of grain legumes. In: Knight R (ed) Linking research and marketing opportunities for pulses in the 21st century. Kluwer Academic Publishing, Dordrecht, The Netherlands, pp 467–480

Cocu N, Conrad K, Harrington R, Rounsevell MDA (2005a) Analysis of spatial patterns at a geographical scale over north-western Europe from point-referenced aphid count data. Bull Entomol Res 95:47–56 http://www.cabi-publishing.org/Journals.asp?PID=16

Cocu N, Harrington R, Hutte M, Rounsevell MDA (2005b) Spatial authocorrelation as a tool for identifying the geographical patterns of aphid annual abundance. Agric For Entomol 7:31–43

Cocu N, Harrington R, Rounsevell MDA, Worner SP, Hullé M (2004) Geographical location, climate and land use influences on the phenology and numbers of the aphid, *Myzus persicae,* in Europe. J Biogeogr 31:1–18

Cooke WC 1963. Ecology of the pea aphid in the Blue Mountain area of eastern Washington and Oregon. U.S. Department of Agriculture, Agricultural Research Service, Technical Bulletin 1287, Washington DC

Dixon AFG (1998) Aphid ecology: an optimization approach. Chapman & Hill, London

Dixon AFG, Kindlmann P (1990) Role of plant abundance in determining the abundance of herbivorous insects. Oecologia 83:281–283

Duetting PS 2002. Effect of field pea surface wax variation on infection of the pea aphid by the fungal pathogen, *Pandora neoaphidis*. M.S. Thesis, University of Idaho, Moscow, Idaho

Dunn JA, Wright DW (1955) Overwintering egg population of the pea aphid in East Anglia. Bull Entomol Res 46:389–392

Eichmann RD (1940) The pea aphid on canning peas in eastern Washington as influenced by alfalfa plantings. J Econ Entomol 33:137–139

Eichmann RD, Webster RL 1940. The influence of alfalfa on the abundance of the pea aphid on peas grown for canning in southeastern Washington. Washington Agricultural Experiment Station Bulletin 389, Pullman, Washington

Eigenbrode SD, White C, Rohde M, Simon CJ (1998) Behavior and effectiveness of adult *Hippodamia convergens* (Coleoptera: Coccinellidae) as a predator of *Acyrthosiphon pisum* on a glossy-wax mutant of *Pisum sativum*. Environ Entomol 27:902–909

Ekbom B (1994) Arthropod predators of the pea aphid, *Acyrthosiphon pisum* Harr. (Hom., Aphididae) in peas (*Pisum sativum* L.), clover (*Trifolium pretense* L.) and alfalfa (*Medicago sativa* L.). J Appl Entomol 117:469–476 http://www.blacksci.co.uk/products/journals/xjae.htm

Elberson LR 1992. Studies of the Russian wheat aphid, *Diuraphis noxia* (Homoptera: Aphididae) and its natural enemies in northern Idaho. M.S. Thesis, University of Idaho, Moscow, Idaho

Evans EW (2004) Habitat displacement of North American ladybirds by an introduced species. Ecology 85:637–647

Feng M-G, Johnson JB, Kish LP (1990) Survey of entomopathogenic fungi naturally infecting cereal aphids (Homoptera: Aphididae) of irrigated grain crops in southwestern Idaho. Environ Entomol 19:1534–1542

Ferrari J, Godfray HCJ, Faulconbridge AS, Prior K, Via S (2006) Population differentiation and genetic variation in host choice among pea aphids from eight host plant genera. Evolution 60:1574–1584

Frazer BD, Gilbert N, Nealis V, Raworth DA (1981) Control of aphid density by a complex of predators. Can Entomol 113:1035–1041

Gutierrez AP, Baumgaertner JU, Summers CG (1984) Multitrophic level models of predator prey energetics. III. A case study in an alfalfa ecosystem. Can Entomol 116:950–963

Hagedorn DJ (ed) (1984) Compendium of pea diseases. American Phytopathological Society, St. Paul, MN

Halfhill JE, Featherston PE (1973) Inundative releases of *Aphidius smithi* against *Acyrthosiphon pisum*. Environ Entomol 2:469–472

Halfhill JE, Featherston PE, Dickie AG (1972) History of the *Praon* and *Aphidius* parasites of the pea aphid in the Pacific Northwest. Environ Entomol 1:402–405

Hampton RO (1983) Pea leaf roll in Northwestern U.S. pea seed production areas. Plant Dis 67:1306–1309 http://www.scisoc.org/pubs/

Hampton RO, Weber KA (1983a) Pea streak virus transmission from alfalfa to peas: virus-aphid and virus-host relationships. Plant Dis 67:305–307

Hampton RO, Weber KA (1983b) Pea streak and alfalfa mosaic viruses in alfalfa: reservoir of viruses infectious to *Pisum* peas. Plant Dis 67:308–310

Harrington R (2002) Insect pests and global environmental change. In: Douglas I (ed) Encyclopedia of global environmental change. John Wiley and Sons, Ltd., Chichester, UK, pp 1–6

Harrington R (2003) Turning up the heat on pests and diseases: a case study for barley yellow dwarf virus. In: The BCPC international congress, crop science and technology: congress proceedings. British Crop Protection Council, Hampshire, UK, pp 1195–1200

Harrington R, Clark SJ, Welham SJ, Verrier PJ, Denholm CH, Hulle M, Maurice D, Rounsevell MD, Cocu N (2007) Environmental change and the phenology of European aphids. Glob Chang Biol 13:1550–1564 http://www.blacksci.co.uk/products/journals/gcb.htm

Harrington R, Fleming RA, Woiwod IP (2001) Climate change impacts on insect management and conservation in temperate regions: can they be predicted?. Agric For Entomol 3:233–240 http://www.blackwellpublishing.com/journal.asp?ref=1461-9555

Hodgson EW (2007) Aphids in alfalfa. Utah Pests Fact Sheet Ent-107-07. Utah State University Extension, Logan, Utah

Homan HW, Stoltz RL, Schotzko DJ (1992) Aphids on peas and lentils and their control. Current information series 748. University of Idaho, Moscow, Idaho

Houghton JT, Ding Y, Griggs DJ, Noguer M, van der Linden PJ, Xiaosu D, Maskell K, Johnson CA (2001) Climate change 2001: the scientific basis. Cambridge University Press, Cambridge

Hutchison W, Hogg DB (1984) Demographic statistics for the pea aphid (Homoptera: Aphididae) in Wisconsin and a comparison with other populations. Environ Entomol 13:1173–1181

Johansen C, Baird C, Ritner R, Fisher G, Undurraga J, Lauderdale R (1979) Alfalfa seed insect pest management. Western regional extension publication 0012. Washington State University, Pullman, Washington

Johansen CA, Eves JD (1973) Development of a pest management program on alfalfa grown for seed. Environ Entomol 2:515–517

Jones PD (1988) Hemispheric surface air temperature variations: recent trends and an update to 1987. J Clim 1:654–660

Kaiser WJ, Ramsey MD, Makkouk KM, Bretag TW, Acikġov N, Kumar J, Nutter FW (2000) Foliar diseases of cool season food legumes and their control. In: Knight R (ed) Linking research and marketing opportunities for pulses in the 21st century. Kluwer Academic Publishers, Dordrecht, pp 437–455

Karley AJ, Parker WE, Pitchford JW, Douglas AE (2004) The mid-season crash in aphid populations: why and how does it occur?. Ecol Entomol 29:383–388 http://www.blacksci.co.uk/products/journals/eent.htm

Klein RE, Larsen RC, Kaiser WJ (1991) Virus epidemic of grain legumes in eastern Washington. Plant Dis 75:1186

Klowden MJ (2002) Physiological systems in insects. Academic Press, Orlando, FL

LaMana MI, Miller JC (1996) Field observations on *Harmonia axyridis* Pallas (Coleoptera: Coccinellidae) in Oregon. Biol Control 6:232–237

Larsen RC, Kaiser WJ, Klein RE (1996) Alfalfa, a non-host of pea enation mosaic virus in Washington State. Can J Plant Sci 76:521–524

Lawton JH (1995) The response of insects to environmental change. In: Harrington R, Stork NE (eds) Insects in a changing environment. Academic Press, London, pp 3–26

Markkula M (1963) Studies on the pea aphid, *Acyrthosiphon pisum* Harris (Hom., Aphididae), with special reference to the difference in the biology of the green and red forms. Ann Agric Fenn Seria Animali Nocentia 2:1–30

McVean RIK, Dixon AFG, Harrington R (1999) Causes of regional and yearly variation in pea aphid numbers in eastern England. J Appl Entomol 123:495–502 http://www.blacksci.co.uk/products/journals/xjae.htm

McWhorter FP, Cook WC (1958) The hosts and strains of pea enation mosaic virus. Plant Dis Rep 42:51–60

Muehlbauer FJ, Kaiser WJ, Clement SL, Summerfield RJ (1995) Production and breeding of lentil. In: Sparks DL (ed) Advances in agronomy. Academic Press, Inc., San Diego, pp 283–332

Muehlbauer FJ, Short RW, Kraft JM (1983) Description and culture of dry peas. US Department of Agriculture, Agricultural Research Service, ARM-W-37, Oakland, CA

Müller FP (1980) Wirtspflanzen, Generationenfolge und reproductive Isolation infraspezifischer Formen von *Acyrthosiphon pisum*. Entomol Exp Appl 28:145–157

Peccoud J, Ollivier A, Plantegenest M, Simon J-C (2009) A continuum of genetic divergence from sympatric host races to species in the pea aphid complex. Proc Natl Acad Sci USA 106: 7495–7500

Pickering J, Gutierrez AP (1991) Differential impact of the pathogen, *Pandora neoaphidis* (R. & H.) Humber (Zygomycetes: Entomophthorales) on the species composition of *Acyrthosiphon* aphids in alfalfa. Can Entomol 123:315–320

Rasmussen LA, Conway H (2004) Climate and glacier variability in western North America. J Climate 17:1804–1815

Rethwisch M (2008) Pink form of pea aphid found for first time in Nebraska alfalfa. Crop Watch, University of Nebraska-Lincoln Extension, Lincoln, Nebraska

Robeson SM (2004) Trends in time-varying percentiles of daily minimum and maximum temperature over North America. Geophys Res Lett 31:L04203 doi:10.1029/2003GL019019

Rockwood LP (1950) Entomogenous fungi of the family Entomophthoraceae in the Pacific Northwest. J Econ Entomol 43:704–707

Rockwood LP, Reeher MM (1943) Forecasting outbreaks of the pea aphid on fall-sown annual legumes in the Pacific Northwest. J Econ Entomol 36:832–837

Rutledge CE, Robinson A, Eigenbrode SD (2003) Effects of a simple plant morphological mutation on the arthropod community and the impacts of predators on a principal insect herbivore. Oecologia 135:39–50

Sandström J (1994) High variation in host adaptation among clones of the pea aphid, *Acyrthosiphon pisum* on peas, *Pisum sativum*. Entomol Exp Appl 71:245–256

Sandström J (1996) Temporal changes in host adaptation in the pea aphid, *Acyrthosiphon pisum*. Ecol Entomol 21:56–62

Skirvin DJ, Perry JN, Harrington R (1997) The effect of climate change on an aphid-coccinellid interaction. Glob Chang Biol 3:1–11 http://www.blacksci.co.uk/products/journals/gcb.htm

Smatas R, Tamosiunas K, Semaskiene R, Dabkevicius Z, Lazauskas S (2008) Trends in aphid occurrence in spring barley in 1976–2007. J Plant Protect Res 48:275–282

Snyder WE, Clevenger GM, Eigenbrode SD (2004) Intraguild predation and successful invasion by introduced beetles. Oecologia 140:559–565

Snyder WE, Ives AR (2003) Interactions between specialist and generalist natural enemies: parasitoids, predators, and pea aphid biocontrol. Ecology 84:91–107

Stoltz RL, Forster RL (1984) Reduction of pea leaf roll of peas (*Pisum sativum*) with systemic insecticides to control the pea aphid (Homoptera: Aphididae) vector. J Econ Entomol 77: 1537–1541

Thacker JI, Thieme T, Dixon AFG (1997) Forecasting of periodic fluctuations in annual abundance of the bean aphid: the role of density dependence and weather. J Appl Entomol 121:137–145

Thackray DJ, Diggle AJ, Berlandier FA, Jones RAC (2004) Forecasting aphid outbreaks and epidemics of cucumber mosaic virus in lupin crops in a Mediterranean-type environment. Virus Res 100:67–82 http://www.elsevier.com/locate/virusres

van den Bosch R, Schlinger EI, Lagace CF, Hall JC (1966) Parasitization of *Acyrthosiphon pisum* by *Aphidius smithi*, a density dependent process in nature (Homoptera: Aphididae) (Hymenoptera: Aphidiidae). Ecology 47:1049–1055

Via S (1991) Specialized host plant performance of pea aphid clones is not altered by experience. Ecology 74:1420–1427

Wellings PW, Dixon AFG (1987) The role of weather and natural enemies in determining aphid outbreaks. In: Barbosa P, Schultz JC (eds) Insect outbreaks. Academic Press, Inc., San Diego, pp 313–346

White C 1998. Effects of *Pisum sativum* surface waxbloom variation on herbivores and predators. M.S. Thesis, University of Idaho, Moscow, Idaho

White C, Eigenbrode SD (2000) Effects of surface wax variation in *Pisum sativum* on herbivorous and entomophagous insects in the field. Environ Entomol 29:773–780

Woiwod IP, Harrington R (1994) Flying in the face of change – the rothamsted insect survey. In: Leigh RA, Johnston AE (eds) Long term research in agricultural and ecological sciences. CABI, Wallingford, UK, pp 321–342

Young FL, Ogg AG, Boerboom CM, Alldredge JR, Papendick RL (1994) Integration of weed management and tillage practices in spring dry pea production. Agron J 86:868–874 http://agron.scijournals.org/

Chapter 8
Temperature, Seasonal Development and Distribution of Insects with Particular Reference to Aphids

Anthony F.G. Dixon and Graham W. Hopkins

Abstract That several species of aphids can feed and coexist on the leaves of certain trees is puzzling. Of these aphids some only grow and reproduce actively in spring and autumn, and others only in summer. This temporal patterning in reproductive activity is well documented and common. In the past the summer reproductive diapause in the sycamore aphid was attributed to the low quality of the food available during summer to this and other aphids living on the leaves of deciduous trees. However, the fact that another species belonging to the same genus feeds on the same leaves and is reproductively active during summer indicates that it is unlikely that reproductive diapause in the sycamore aphid is mainly a response to poor food quality. The seasonal trend in temperature is the opposite of that in nutrition, low in spring and autumn and high in summer. A study of five species of aphids coexisting on the leaves of birch indicates that the temporal patterns in their reproductive activity are associated with differences in their thermal tolerances. The thermal tolerance range of each species, their thermal window, is approximately 20°C, with some species better adapted to and therefore reproductively active when temperatures are low in spring and autumn and other species when temperatures are high in summer. Thermal windows of 20°C are not specific to aphids as an analysis of similar data for 66 species from eight orders of insects indicates it is a general phenomenon. In addition, temperate species of aphid, like other poikilotherms, do better at low temperatures than subtropical and tropical species and vice versa. This indicates that it is more likely that a physiological constraint, their thermal windows, determines the seasonal dynamics and distribution of aphids, and also enables several species to coexist.

Keywords Aphids · Coexistence · Distribution · Food quality · Resource partitioning · Seasonal dynamics · Temperature · Thermal tolerance

---

A.F.G. Dixon (✉)
Department of Theoretical Ecology, Institute of Systems Biology and Ecology AS CR, CZ-37005 České Budějovice, Czech Republic; University of East Anglia, Norwich, UK
e-mail: a.f.dixon@uea.ac.uk

P. Kindlmann et al. (eds.), *Aphid Biodiversity under Environmental Change*,
DOI 10.1007/978-90-481-8601-3_8, © Springer Science+Business Media B.V. 2010

8.1 Introduction

Temperature is a very important component of the environment of poikilothermic organisms as it determines both their rate of development and survival. For each species there is a range of temperatures over which it can survive, and within this range there is a more limited range over which it can grow and reproduce. The effect of temperature on growth and reproduction is well studied, especially of pest insects in temperate regions. Most of the studies have been on determining the lower developmental threshold and sum of effective temperatures needed for development as these are seen as useful for predicting when in a year a pest will become active. In addition, it is of interest to know whether a pest can survive low winter temperatures and quickly become active again when temperatures increase in spring. That is, the information falls into two categories: (1) The effect of temperature on growth and reproduction, mainly measured in terms of its effect on rate of development covering the range of temperatures over which each organism can develop, i.e., its ecologically relevant temperature range; (2) The effect of extremes of temperature on survival. Species can survive temperatures outside their ecologically relevant temperature range, and have evolved many ways of anticipating and surviving extreme temperatures.

In temperate regions insects are not all equally active throughout the year. For example, of the aphids that over-winter and live mainly on deciduous trees, some appear very early in spring, complete 2–3 generations and then produce eggs, which remain dormant until the following spring. That is, they spend something like 9 months of each year dormant in the egg stage. Others appear early, actively grow and reproduce in spring and autumn, and although present in summer either do not grow or reproduce and enter diapause, or they migrate, colonize and reproduce on herbaceous plants in summer and return to their woody host in autumn, i.e., they host alternate. Other species appear much later and actively grow and reproduce all through summer and produce eggs in early autumn. Interestingly species of aphids showing one or other of all these life cycles can occur on the same species of tree often even on the same tree. In the past these life cycles have been viewed mainly as a means of resource partitioning based on adaptations to the seasonal development of plants (Dixon 1998). The suggestion that this resource partitioning might be more associated with adaptation of individual species to particular ranges of the wide seasonal range in temperature conditions (Dixon 1973), was ignored mainly because of the absence of a mechanism by which this could be achieved.

Poikilothermic organisms in the tropics tend to have higher threshold temperatures for development than those in temperate regions (Honěk 1996; Trudgill 1995; Trudgill et al., 2005). Similarly, this indicates that like seasonal development, distribution may also be associated with adaptations to particular temperature ranges.

In terms of the effect of temperature on development aphids are a particularly well-studied group. The objective of this chapter is to show that there is a pattern in the effect of temperature on their development that is associated with their seasonal development and distribution.

8.2 Thermal Time

In poikilothermic organisms development rates increase with temperature following a sigmoidal curve if measured over the ecologically relevant range of temperatures. At the lower temperatures the rate tends to level off making it difficult to define a point when development ceases. A surrogate value for the lower developmental threshold, which has been widely adopted, is to use the value obtained by extrapolating the linear portion of the relationship between rate of development and temperature back to intercept the X-axis (Fig. 8.1). This virtual value is referred to as the lower developmental threshold (LDT) or basal temperature (t_b). At the higher temperatures, increasing heat stress gradually reduces the slope of the curve, which reaches a maximum development rate before decreasing due to heat stress. The point of the maximum development rate is referred to as the optimum temperature (t_0) (Trudgill et al., 2005) or maximum temperature (t_{dmax}) (Birkemoe and Leinaas 2000). Optimum implies it is the most favourable temperature for development. However, organisms are subject to fluctuations in temperature and are unlikely to be adapted to doing best at one particular temperature but to optimize their performance over the range of temperatures most frequently experienced in the field.

Therefore, the term t_{dmax} is preferred to t_0 and for conformity t_{dmin} to t_b, for the temperatures at which development rate is fastest and zero, respectively. Like t_{dmin}, t_{dmax} is also not a precise temperature but has to be estimated by extrapolating the linear increasing and decreasing parts of the relationship between development rate and temperature, or alternatively fitting a curve to the upper part of the relationship (Dixon et al., 2009; Lamb et al., 1987; Liu and Meng 1989; Summers et al., 1984). The point of intersect or the uppermost point of the curve is t_{dmax}.

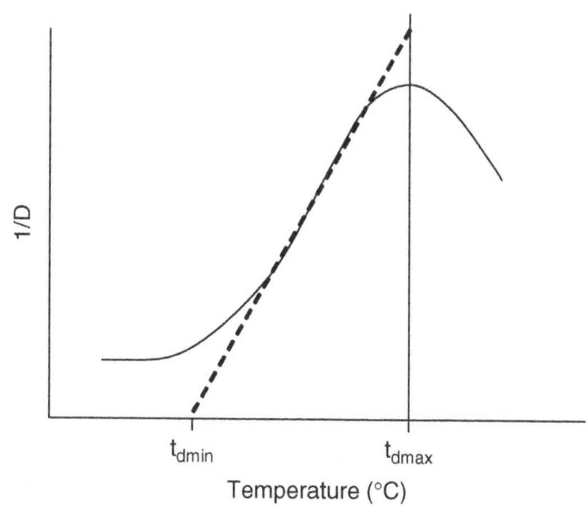

Fig. 8.1 Relationship between temperature and rate of development (1/D) for poikilothermic organisms (t_{dmin} and t_{dmax} indicate the lower and upper developmental thresholds)

8.2.1 Aphid Biology

The data for this analysis comes mainly from a study of the seasonal development and rate of development and growth of five species of aphids living on birch, *Betula* spp. (Hopkins 1996) and other information scattered in the literature.

8.3 Seasonal Development in Aphids

8.3.1 Aphids on Sycamore

The most important aphid on sycamore, *Acer pseudoplatanus,* is the sycamore aphid, *Drepanosiphum platanoides*, which hatches just before bud burst of its host, initially feeds on the expanding buds and then colonizes leaves equally throughout the canopy and thrives on growing leaves. On the approach of summer, as the leaves stop growing, the second generation adults present at that time vacate the tops of the trees and spend the summer in reproductive diapause in the lower canopy. At the onset of autumn these aphids start reproducing and gradually recolonize the upper parts of the canopy. This behaviour is associated with dramatic changes in the nutritional quality of the leaves for the sycamore aphid, which is high in spring, low in summer and high again in autumn. In addition, the temperature is low in spring, high in summer and low in autumn. That is, both food quality and temperature change dramatically during the course of a season (Fig. 8.2).

As indicated in the Introduction the life history of the sycamore aphid was thought to be driven by the dramatic changes in food quality, which is high when the leaves are actively growing and senescing in spring and autumn, respectively, and poor when the leaves are mature in summer (Mordvilko 1908). In addition, during

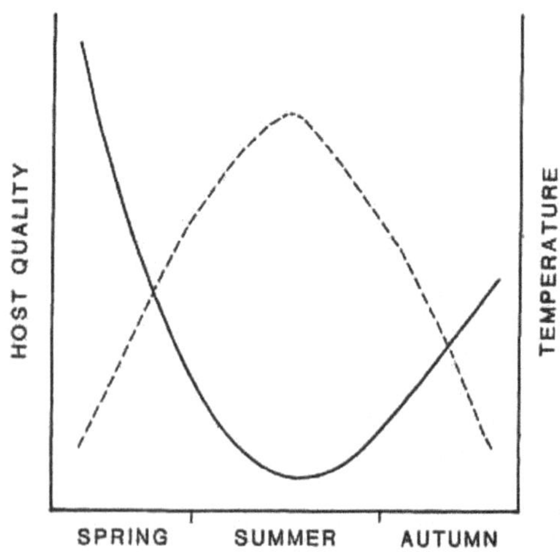

Fig. 8.2 Seasonal change in host quality (*solid line*) and temperature (*dashed line*) experienced by aphids living on trees

cool periods in summer there is a tendency for the aphids to move higher up in the canopy. This tends to indicate the aphid could be experiencing heat stress at this time of the year. One piece of evidence for this comes from changes in coloration shown by the sycamore aphid during the course of a year. In spring and autumn, when temperatures are low, each abdominal segment of the aphids maturing then has a melanic band on its dorsal, lateral and ventral surfaces. In addition, the head, thorax and legs are also darker in colour. At these times of the year the aphids appear quite black and are more conspicuous than they are in summer. The latter maturing first generation and early maturing second generation adults have fewer and smaller and less darkly pigmented melanic bands, and generally are a paler colour. Those maturing in summer have no abdominal bands and are generally a very pale colour and inconspicuous (Fig. 8.3).

During the summer there can also be a marked difference in the temperature of the leaves in the upper and lower canopy. In the middle of the day it can be as much as 10°C warmer in the upper canopy, reaching temperatures as high as 32°C (Dixon 2005). Although an aphid's body temperature is usually close to ambient, it can, by moving onto the upper surface of leaves in autumn or avoiding leaves exposed to the sun in summer, increase or decrease the heating effect of absorbing solar radiation. The presence of melanic pigmentation similarly can increase the absorption of solar radiation and increase body temperature by as much as 2°C (Dixon 1972). Thus, it is possible that this aphid attempts to reduce heat stress in summer by reducing its pigmentation and occupying shaded leaves in the lower canopy, and increases its body temperature in spring and autumn by increasing its pigmentation and behavioural changes that result in greater exposure to solar radiation. Thus, temperature could be a more important factor determining this aphid's life cycle than previously thought.

Of the other aphids that live on sycamore, *Periphyllus testudinaceus* hatches very early in the year and completes one generation before the buds break and their

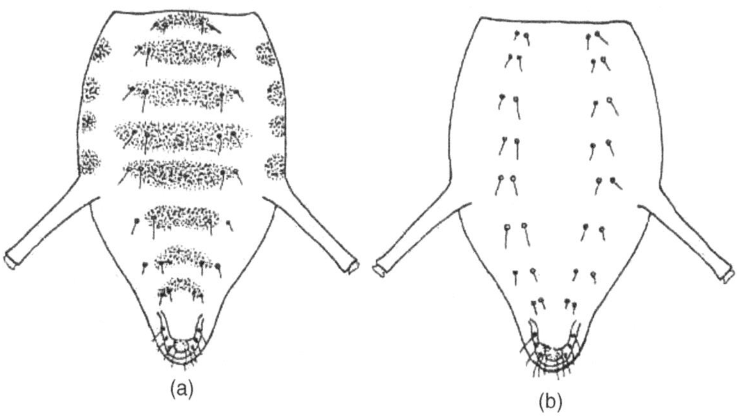

Fig. 8.3 The pattern of melanic pigmentation on the dorsum of the abdomen of sycamore aphids maturing in spring or autumn (**a**) and in summer (**b**)

offspring, which colonize and mature on the emerging leaves, give birth to very specialized larvae that do not develop but aestivate as first instar larvae. Late in the year, when the leaves start to yellow these larvae resume development and their offspring are often laying over-wintering eggs well after leaf fall. The third species, *Drepanosiphum acerinum*, hatches well after bud burst and at a time when the leaves are nearly fully grown. It continues to reproduce and develop throughout summer and produces sexuals and over-wintering eggs quite early in autumn. That is, the two species that hatch early, *P. testudinaceus* and *D. platanoides*, thrive on sycamore early and late in the year and spend the summer in aestivation. The species that hatches last, *D. acerinum*, thrives and continues to reproduce throughout summer, and produces sexuals and over-wintering eggs early in autumn. This seems to indicate that it is not poor nutrition but mainly high temperatures that make sycamore a poor host for *P. testudinaceous* and *D. platanoides* in summer. In addition, the sequence of hatching of these three species implies that *P. testudinaceus* has a lower t_{dmin} than *D. platanoides*, which has a lower t_{dmin} than *D. acerinum*, and the periods these three species spend in aestivation is inversely associated with their average time of egg hatch.

8.3.2 Aphids on Birch

There are more than three species of aphids on birch (*Betula* spp.). Of these the seasonal development and rate of development of *Betulaphis quadrituberculata, Callipterinella tuberculata, Euceraphis betulae, Kallistaphis flava* and *Monaphis antennata* has been studied in detail. Like on sycamore, one of these aphids, *E. betulae,* hatches early and goes into reproductive diapause in summer and another, *C. tuberculata*, hatches late and continues reproducing throughout summer. The other three species hatch at definite times between these two extremes. That is, as on sycamore there are species of birch aphid that hatch in a definite sequence, which can be related to their subsequent behaviour. However, in this case there is data on the rate of development over a wide range of temperatures and consequently good estimates of the t_{dmin} for each species. An analysis of the relationship between the time to egg hatch from the beginning of the year and t_{dmin} for these five species gives:

$$\text{Time to egg hatch} = 21.01 + 24.29 \log t_{dmin} \tag{1}$$

Although this relationship is not significant (r=0.85; n=5; 0.05<P<0.1) it lends strong support to the suggestion that the time of egg hatch of each species is determined by its t_{dmin}.

Plotting rate of development (1/*D*) against temperature for *C. tuberculata* and *E. betulae* reveals that *E. betulae* does better than *C. tuberculata* at low temperatures and vice versa at high temperatures. Interestingly, the t_{dmax} of *E. betulae* is lower than that of *C. tuberculata,* and by about the same number of degrees

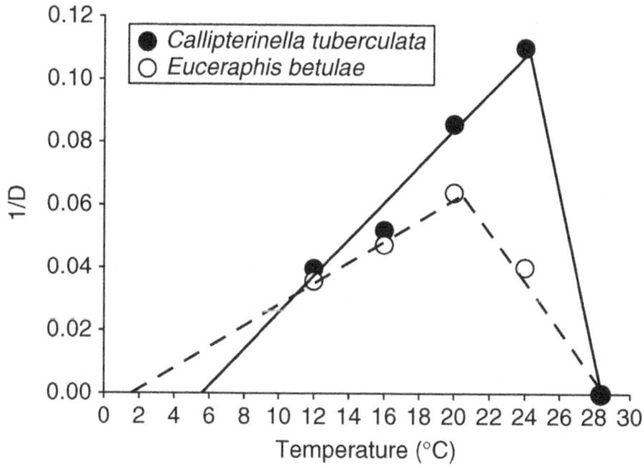

Fig. 8.4 Relationship between the rate of development (1/D) of *Euceraphis betulae* and *Callipterinella tuberculata* and temperature

as the difference in their lower developmental thresholds (Fig. 8.4). This point will be returned to and discussed in greater detail under distribution. Using the data available on the population growth rate at different temperatures rather than the rate of development reveals, not unexpectedly, the same trends in performance with temperature for these two species (Fig. 8.5). This suggests that the species on birch, like those on sycamore, are adapted either more to the temperature conditions prevailing in spring and autumn (e.g., *E. betulae*) or summer (e.g., *C. tuberculata*).

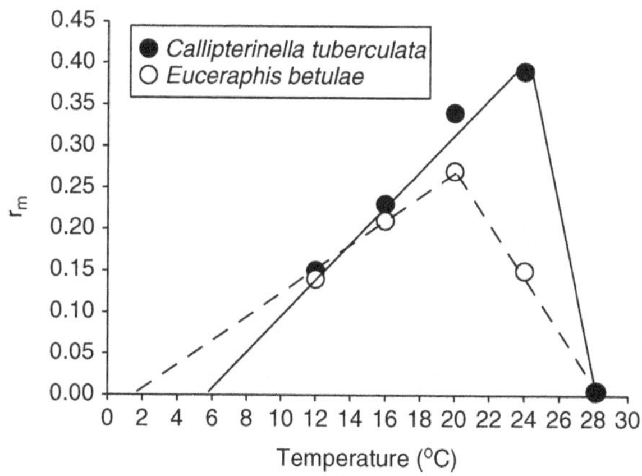

Fig. 8.5 Relationship between population growth rate of *Euceraphis betulae* and *Callipterinella tuberculata* and temperature

8.3.3 Host Alternating Aphids

Host alternating aphids also tend to hatch early on their woody primary hosts, but then colonize herbaceous hosts during summer returning to their woody hosts in autumn. That is, in this case a low t_{dmin} does not prevent them from thriving in summer on herbaceous plants. This is possibly because herbaceous plants at that time are a high quality source of food (Kundu and Dixon 1995). That is, both food quality and temperature are important for aphids. Their size, which affects their fecundity, is a consequence of the relative effect of food quality and temperature on their growth and developmental rates. Both an increase in food quality and temperature results in an increase in the growth rate measured as the increase in weight per unit weight per unit time. However, increases in temperature disproportionately decrease the time it takes to reach maturity. Similarly, the time to maturity is affected by an increase in food quality, but the effect is subproportional. As a consequence aphids are small when reared at high temperatures and large when reared on high quality food and vice versa (Dixon 1998). In exploiting the complementary growth patterns of herbaceous and woody plants host-alternating aphids can within certain limits overcome the constraints imposed by the seasonal changes in temperature. That is, host alternation is one way of exploiting a resource that varies seasonally. The effect of temperature on the rate of development has been important in determining the outcome, whether it be host alternation or limiting development to certain periods during the vegetative period. For simplicity, however, the emphasis in this chapter is mainly on the role of temperature, but seasonal changes in nutrition have also been important.

8.4 Seasonal Development in Predators of Aphids

The duration of the increase, peak and decline in the abundance of the immature aphids on plants each year is both seasonal and ephemeral. These temporary resources are exploited by a sequence of aphidophagous insect predators. The temporal sequence in the appearance of the immature stages of ladybirds (coccinellids) and hoverflies (syrphids) is consistent from year to year, with the hoverfly larvae appearing first and then the ladybird larvae In addition, although the autumnal peak in abundance of the sycamore aphid is on average larger than the spring peak, and is attacked by more hoverfly larvae, it is not exploited by ladybird larvae. These temporal patterns in the attack sequence are associated with a difference in the lower developmental thresholds of these two groups of predators. The t_{dmin} of hoverflies (4°C on average) enables them to be active at lower temperatures and to develop faster between 10 and 27°C than ladybirds, whose t_{dmin} is 10°C on average (Fig. 8.6). As a consequence, in spring, when temperatures are low but increasing, hoverflies appear before and complete their development more quickly than ladybirds, and in autumn, when temperatures are generally lower than in summer and decreasing, only hoverflies are likely to complete their development before the

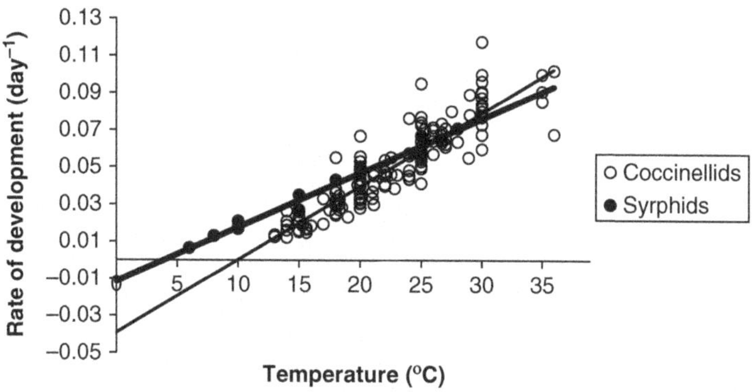

Fig. 8.6 Relationship between the rate of development (D–[1]) and temperature of aphid eating coccinellids and syrphids (after Dixon et al. 2005). The data points are for several species of hoverflies and ladybirds

aphid disappears (Dixon et al., 2005). That is, for predators of aphids there is also good evidence to indicate that their appearance and seasonal activity patterns are determined by their t_{dmin}.

8.5 Distribution of Birch Aphids in Europe

Comparison of the rates of development of temperate and tropical species of poikilotherms belonging to the same genus indicate that temperate species become active and develop faster at low temperatures than tropical species and vice versa for the rate of development at high temperatures. However, the differences in t_{dmin} of temperate and tropical species are similar to those recorded for early and late hatching aphids living on the same tree in northern Europe. Thus, in the case of *E. betulae* and *C. tuberculata*, which live on a very widely distributed plant, birch, the prediction would be that *E. betulae* would occur predominantly in the north of Europe and *C. tuberculata* in the south. Plotting the locality records for these two species in Europe reveals very similar distributions (Figs. 8.7 and 8.8). The fewer records for *C. tuberculata* can be attributed to it being less common than *E. betulae.* Their similar distributions are more clearly revealed by plotting the distribution of *C. tuberculata* within the climatic envelop of *E. betulae,* defined in terms of the mean February and July temperatures at the localities where each of these two species are recorded (Fig. 8.9). The distribution of *C. tuberculata* falls within the climatic envelope of *E. betulae*, and there is no apparent tendency for *C. tuberculata* to be more frequently recorded in more southerly localities, i.e., those with warmer summers and winters.

In summary, these two species have very similar distributions despite the differences in their t_{dmin}. For these species the differences in t_{dmin} are associated with seasonal development not their distribution within Europe.

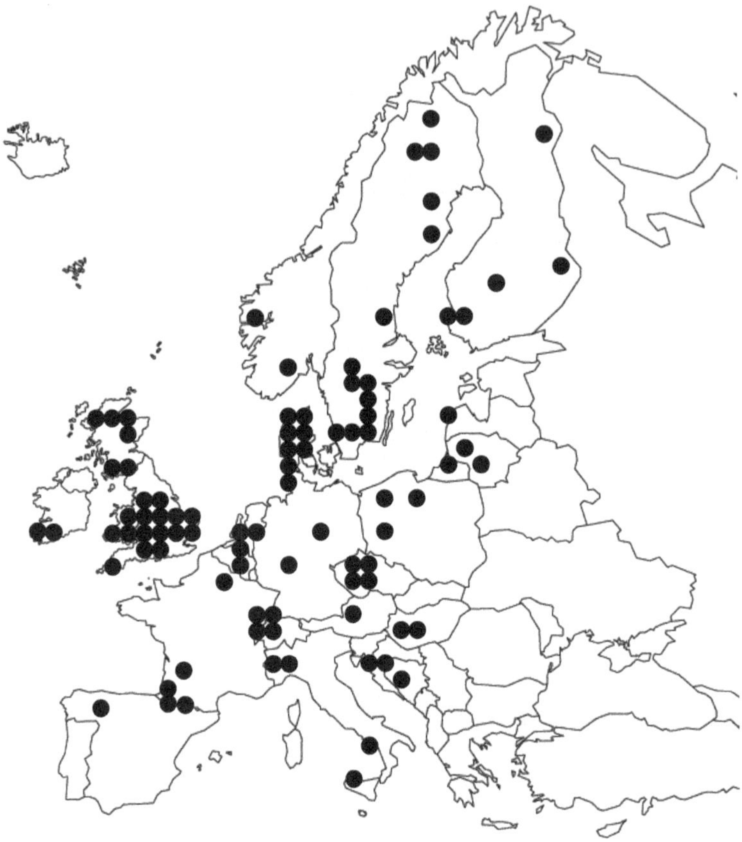

Fig. 8.7 Distribution of *Euceraphis betulae* in Europe

8.6 World Wide Distribution of Aphids

A plot of the rate of development on temperature for a temperate species of aphid, *E. betulae*, and a tropical species, the soybean aphid, *Aphis glycines,* reveal a very marked difference in their t_{dmin} and possibly more importantly in their t_{dmax} (Fig. 8.10). This clearly indicates that *E. betulae* can develop better at low temperatures than *A. glycines* and vice versa at high temperatures. In addition, the t_{dmax} of *A. glycines* (27°C) is considerably higher than that of *E. betulae* (20°C). Interestingly, $t_{dmax} - t_{dmin} \approx 18°C$ for both these species, but the ranges are different 9–27°C for *A. glycines* and 1–20°C for *E. betulae.* Disappointingly, the performance of organisms at the higher temperatures at which one would expect them to show heat stress is rarely recorded (e.g., Fig. 2 in Trudgill et al., 2005). This is also true of aphids and if there is such data it usually only includes performance at one temperature greater than t_{dmax}. Therefore, it is often more difficult to obtain a good estimate of t_{dmax} than t_{dmin}. However, such data is available for 24 species of aphids

Fig. 8.8 Distribution of *Callipterinella tuberculata* in Europe

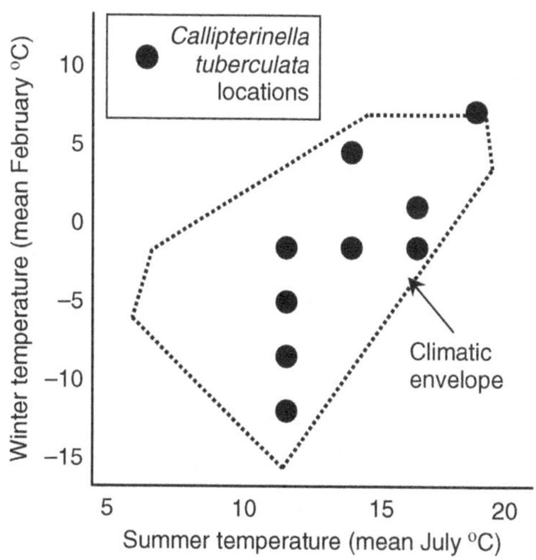

Fig. 8.9 Distribution of *Callipterinella tuberculata* plotted within the climatic envelop of *Euceraphis betulae* defined in terms of the mean winter and summer temperatures at the localities where each species are recorded

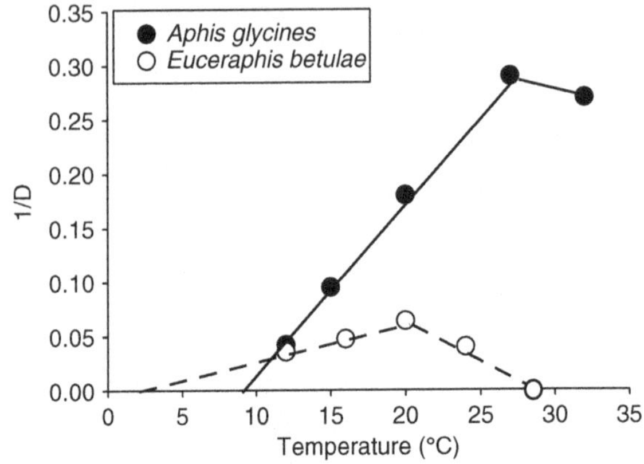

Fig. 8.10 Relationship between the rate of development (1/D) of *Euceraphis betulae* and *Aphis glycines*, and temperature

and indicates that t_{dmax} increases with t_{dmin} and although there is a lot of variability it indicates that change in these two values could be directly proportional – a one degree centigrade increase in t_{dmin} is associated with a similar change in t_{dmax} (Fig. 8.11).

Thus although the little information is of poor quality it nevertheless supports the contention that the ecologically relevant range of temperatures for each species spans approximately the same number of degrees (≈18°C), with the temperate species better adapted to operate at lower temperatures than tropical species and vice versa at high temperatures. It is possible that it is more important that tropical species are better able to tolerate high temperatures than develop at low temperatures.

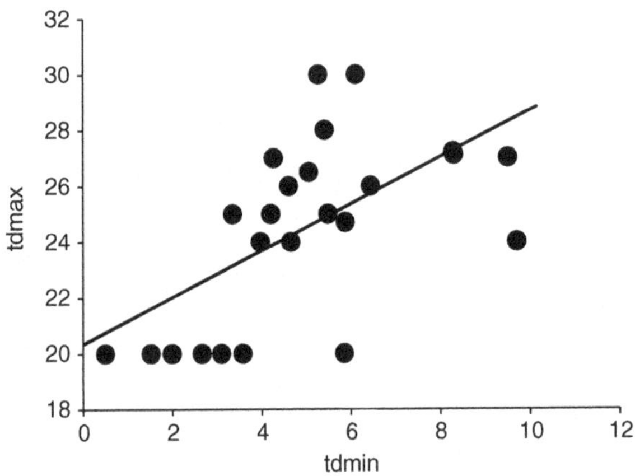

Fig. 8.11 Relationship between t_{dmax} and t_{dmin} for 24 species of aphid (data from Bieri et al. 1983; Bonnemaison 1951; Carroll and Hoyt 1980; El Din 1976; Elliott and Kieckhefer 1989; Hirano et al. 1996; Hopkins 1996; Komazaki 1982; Lamb 1992; Liu and Hughes 1987; Liu and Meng 1989; Liu and Yue 2000; Nowieski et al. 1983; Reed and Semntner 1991; Rohitha and Penman 1983; Watson 1983; Xia 1997; Yamaguchi 1976) ($t_{dmax} = 20.31 + 0.86\ t_{do}$; $r = 0.64$; $n = 24$; $P<0.01$)

8.7 Discussion

The effect of temperature on the development and growth of poikilotherms has been studied over a very long period and is extremely well reported. Even nearly 300 years ago Réaumur (1735, 1736) was aware that there were temperatures below which organisms cannot develop and appreciated that the temperature sum required for complete growth of an organism is a constant. Two hundred years later Bodenheimer (1927), after studying two species of weevil that infest stored grain, *Sitophilus granaria* and *Sitophilus oryzae*, which in the field have a northern and southern distribution, respectively, argued their distribution could be attributed to *S. granaria* doing better at low temperatures than *S. oryzae* and vice versa at high temperatures. That is, nearly a 100 years ago two basic aspects of the relationship between poikilothermic species and their thermal environment had been identified, the constancy of the temperature sum required by each organism for development, now usually measured in day degrees (D°), and that northern and southern species differ in their performance at high and low temperatures, with the northern species doing better at low temperatures than the southern species and vice versa at high temperatures. Since then there have been many studies on survival at low temperatures, possibly driven by an interest in predicting when pest species will become active and of the survival of poikilothermic species in the Arctic and Antarctic. These studies have, not surprisingly, indicated that species can survive the low temperatures they are likely to experience and have evolved a number of ways of doing this (see reviews of Bale 2002 and Hodkinson 2005). Although interesting, these studies have not contributed anything towards defining the mechanism underlying the relationship between rate of development and temperature, mainly because all the indications are that survival at extreme temperatures involves different physiological mechanisms.

Fitness is often viewed in terms of potential rate of population increase, mainly because it is easy to measure. If it is a major component of fitness, and development rate and population rate of increase are correlated, then it would appear to be advantageous for tropical species and those temperate species that develop in summer to have lower developmental thresholds. However, this assumes there are no costs associated with developing even faster at the temperatures prevailing in the tropics and high summer in temperate regions. The study reported here may have identified the cost, their ability to survive at high temperatures. That is, fitness is more constrained by physiological rather than ecological considerations, with the optimization of fitness in a variable thermal environment the central issue. Viewed in this way seasonal development and distribution are very similar in that they both involve adaptation to development over a particular temperature range. Northern species and those temperate species that start developing early in a year, have to be able to tolerate relatively low temperatures, those temperate species that develop in summer and tropical species, relatively high temperatures. This study of aphids indicates that each species can only develop over a relatively narrow range of temperatures (18°) and this holds for both seasonal development and worldwide distribution. In specific terms species that have a low t_{dmin} have a low t_{dmax} and those that have a high t_{dmin} have a high t_{dmax}, and the difference between the two values in each case is similar.

It is now important to consider whether the direct proportional relationship between t_{dmin} and t_{dmax} is a widespread phenomenon or specific to aphids? Studies on Arctic collembola and mites also show that they are well adapted to cold conditions and have a relatively low t_{dmax} (Burn 1984; Convey 1994) and whether they live sheltered under stones or mainly exposed to high temperatures on the surface of the soil determines the extent to which they are adapted to low temperature conditions (Birkemoe and Leinaas 2000; van Straalen 1994). There is a lot of data that indicates that tropical species differ from temperate species in their t_{dmin} and in the slope of the relationship between development rate (1/D) and temperature over the range t_{dmin} to t_{dmax} (Trudgill et al., 2005; Honěk and Kocourek 1990; Trudgill and Perry 1994). However, the relationship between t_{dmax} and t_{dmin} has not previously been considered. This is possibly mainly due to the very few studies that record the effect of high temperatures on development. Thus, it is likely the observation that species can develop only over a very limited range of temperatures is a general phenomenon.

Further support comes from theoretical studies (Charnov and Gillooly 2003; Gillooly et al., 2002), which suggest that the width of the thermal window for each ectothermic species should be about 20°C. As this prediction is based on a thermodynamic model of development then the mechanism is more likely to be physiological than ecological, and might be independent of phylogentic relationships. A metanalysis of data on the thermal requirements for development of 66 species from eight orders of insects obtained from the literature indicate that each species can only develop over a limited range of temperatures – its thermal window (19.8°C) – independent of species traits and phylogeny (Dixon et al., 2009).

Recently climate change and in particular its effect on organisms has become a major issue (Dewar and Watt 1992; Harrington et al., 1999; Watt and Woiwood 1999). In insects the threat is seen in terms of the disruption of the phenological synchrony between host plants and herbivores, hosts and parasitoids, and prey and predators. This issue mainly involves the way these organisms track the seasons and the synchrony between important events in their life histories and that of their hosts/prey and the way this is likely to be disrupted by global warming. Predictions of models that attempt to determine the effect of climate change on the abundance of an insect, which do not include the mechanisms by which the insect's abundance is regulated and how selection is likely to shape its life history strategy, are unlikely to be realistic. For tree-dwelling aphids, in which the processes acting at both the individual and population levels are fairly well understood, Dixon (2003) argues that the evidence indicates the adverse effect of climate change is unlikely to be as threatening as is so often suggested. Climate change induced asynchrony in the frequency distribution in the time of bud burst and egg hatch in any 1 year is likely to be slight and any negative effect on aphid numbers will be compensated for within the year due to the dynamics of the system. In addition, proportionally more of the survivors will be individuals with inherited responses that enable them to track more closely the phenology of their host plant. These are likely to mate with one another and as a consequence the following year proportionally more of the eggs hatch in synchrony with bud burst. That is, coupled with the consequences of population regulation, selection is likely to quickly correct any asynchrony between egg hatch and bud burst. The variability in the time of egg hatch of the different aphids on birch and sycamore, and other aphids (Dixon 1976; Komazaki 1986) and the variability between clones in the response of the development rate to temperature in the pea aphid (*Acyrthosiphon pisum*) (Lamb et al., 1987) indicate that variability in life history traits in aphids is a general phenomenon and sufficient for quickly adapting via natural selection to environmental change.

The prediction is that with global warming average temperatures will increase. Thus, if the distribution of aphids and other organisms is limited, among other things, by the range of temperatures over which each species can develop then it is likely that the distribution of many species will move northwards as temperatures rise. This could occur very rapidly for those species of aphids for which suitable hosts are already present further north. *Hoplocallis pictus,* which characteristically infests oak around the Mediterranean, has recently been recorded in Northern Europe, as has *Aphis solanella* on spindle, *Aphis spiraecola* on apple, *Brachycaudus divaricatae* on plum and *Dysaphis pyri* on pear (Rakauskas 2004). In this context it is interesting to consider the distribution in Europe of the closely related host alternating *Aphis fabae cirsiiacanthoides, A. fabae fabae* and *A. solanella* (Fig. 8.12).

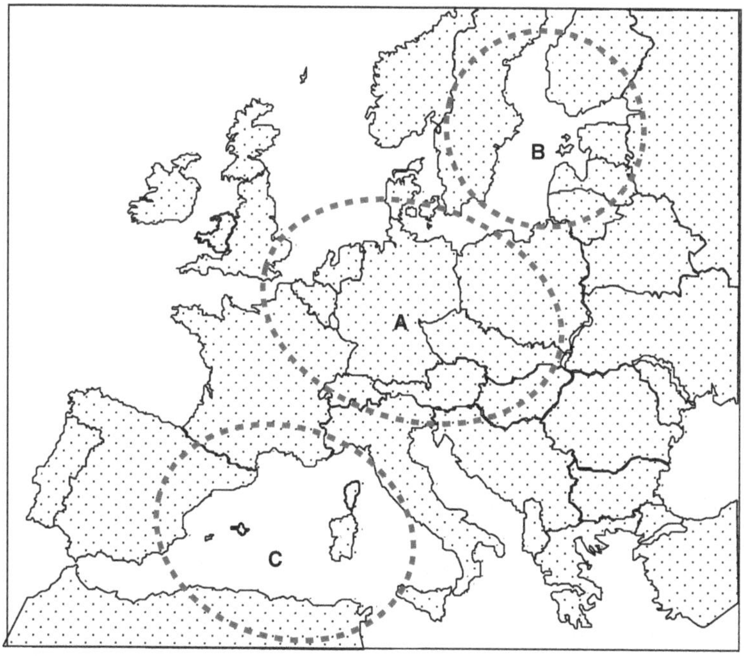

Fig. 8.12 Areas of dominance of (*A*) *Aphis fabae fabae*, (*B*) *Aphis fabae cirsiiacanthoidis* and (*C*) *Aphis solanella* in Europe and North Africa

They all share the same primary hosts but have slightly different ranges of secondary hosts. *A. fabae cirsiiacanthoides* is predominantly a northern species, *A. fabae fabae* a middle European species and *A. solanella* a southern species (Thieme and Dixon 2004). Of these species *A. fabae cirsiiacanthoides* has the lowest t_{dmin} (1.7°C), *A.fabae* a slightly higher t_{dmin} (3.7°C) and *A. solanella* the highest t_{dmin} (8.4°C). Interestingly, *A. solanella* has recently been reported for the first time in Northern Europe and from the far north of Germany (Rakauskas 2004; Thieme, personal communication). However, where the host plant distribution has to shift northwards as well, the expectation is that it will take longer. Many southern species of plants are already invading Northern Europe and when abundant will inevitably be colonized by their specific aphids. Thus, although global warming may not be as a great threat to the abundance of aphids as previously thought it will affect their distribution.

Biochemists at the end of the nineteenth century developed temperature – coefficient equations, based on thermodynamic reasoning for simple inorganic systems, for the effect of temperature on the rate of biological processes (Arrhenius 1889; van't Hoff 1894). It was also appreciated that these relationships do not apply to the entire temperature range, but only narrow segments, depending on the species of animal, or in particular the enzyme(s) that control development. As the temperature rises, the enzymes begin to denature, with the effect that the accelerating effect of temperature on the speed of development may be counteracted by the inactivating effect on the limiting enzyme(s) (Chick and Martin 1910). It is possible limiting enzymes are similarly affected by temperatures at the lower end of the ecologically relevant range. More recently it has been argued that the apparent trade off between t_{dmin} and the D° required for development has a basis in the thermal adaptation of enzymes (van Straalen 1994; van der Have and de Jong 1996; Trudgill et al., 2005). The suggestion that each species of aphid may only be able to develop over a very limited range of temperatures (18°) may help identify the precise nature of the physiological mechanism underlying the seasonal development and distribution of aphids and other poikilotherms. This could be an important step in achieving a better understanding of how communities work and integrating physiology and ecology at the community scale.

Acknowledgements It is a pleasure to thank Jonathan Thacker for helpful comments on this paper. AFGD thanks the Leverhulme Trust for the award of an Emeritus Fellowship that made this study possible.

References

Arrhenius S (1889) Über die Reaktionsgeschwindigkeit bei Inversion von Rohrzucker durch Sauren. Zeitschrift für Physikalische Chemie 4:226–248

Bale JS (2002) Insects at low temperatures: from molecular biology to distributions and abundance. Philosphical Tran R Soc Lond B 357:849–862

Bieri M, Baumgärtner J, Bianchi G, Delucchi V, von Arx R (1983) Development and fecundity of pea aphid (*Acyrthosiphonn pisum* Harris) as affected by constant temperatures and by pea varieties. Mitteilungen der Schweizerischen Entomogischen Gesellschaft 56:163–171

Birkemoe T, Leinaas HP (2000) The effects of temperature on the development of an arctic Collembola (*Hypogastrura tullbergi*). Funct Ecol 14:693–700

Bodenheimer FS (1927) Cited in Chapman, R.N. 1931. Animal ecology. McGraw Hill, New York

Bonnemaison L (1951) Contribution à l'étude des facteurs provoquant l'apparition des formes ailées et sexuées chez leas Aphidinae. Ph.D. Thesis, Paris University

Burn AJ (1984) Life cycle strategies in two Antarctic Collembola. Oecologia 64:223–229

Carroll DP, Hoyt SC (1980) Developmental rate, weight, and ovarian parameters of apple aphids (Homoptera: Aphididae) reared at constant and alternating temperatures. Melandria 44:11–17

Charnov EL, Gillooly JF (2003) Thermal time: body size, food quality and the 10°C rule. Evol Ecol Res 5:43–51

Chick H, Martin CJ (1910) On the "heat coagulation" of proteins. J Physiol 40:404–430

Convey P (1994) Growth and survival strategy of the Antarctic mite *Alaskozetes antarcticus*. Ecography 17:97–107

Dewar RC, Watt AD (1992) Predicted changes in the synchrony of larval emergence and budburst under climatic warming. Oecologia 89:557–559

Dixon AFG (1972) Control and significance of the seasonal development of colour forms in the sycamore aphid, *Drepanosiphum platanoides* (Schr). J Anim Ecol 41:689–697

Dixon AFG (1973) Metabolic acclimatization to seasonal changes in temperature in the sycamore aphid, *Drepanosiphum platanoides* (Schr.) and lime aphid, *Eucallipterus tiliae* L. Oecologia 13:205–210

Dixon AFG (1976) Timing of egg hatch and viability of the sycamore aphid, *Drepanosiphum platanoidis* (Schr.), at bud burst of sycamore, *Acer pseudoplatanus* L. J Anim Ecol 45:593–603

Dixon AFG (1998) Aphid ecology: an optimization approach, 2nd edn. Chapman and Hall, London

Dixon AFG (2003) Climate change and phenological asynchrony. Ecol Entomol 28:380–381

Dixon AFG (2005) Insect herbivore – host dynamics: tree-dwelling aphids. Cambridge University Press, Cambridge

Dixon AFG, Honěk A, Keil P, Ali M, Kotela A, Šizling AL, Jarošík V (2009) Relationship between the minimum and maximum temperature thresholds for development in insects. Funct Ecol 23:257–264

Dixon AFG, Jarošík V, Honěk A (2005) Thermal requirements for development and resource partitioning in aphidophagous guilds. Eur J Entomol 102:407–411

El Din NS (1976) Effects of temperature on the aphid, *Myzus persicae* (Sulz.), with special reference to critically low and high temperature. Zeitschrift für angewandte Entomologie 80:7–14

Elliott NC, Kieckhefer RW (1989) Effects of constant and fluctuating temperatures on immature development and age specific life-tables of *Rhopalosiphum padi* L. (Homoptera, Aphididae). Can Entomol 121:131–140

Gillooly JF, Charnov EL, West GB, Savage VM, Brown JH (2002) Effects of size and temperature on developmental time. Nature 417:70–73

Harrington R, Woiwood I, Sparks T (1999) Climate change and trophic interactions. TREE 14:146–150

van der Have TM, de Jong G (1996) Adult size in ectotherms: temperature effects on growth and differentiation. J Theor Biol 185:329–340

Hirano K, Honda K-I, Miyai S (1996) Effects of temperature on development, longevity and reproduction of the soybean aphid, *Aphis glycines* (Homoptera: Aphididae). Appl Entomol Zool 31:178–180

Hodkinson ID (2005) Adaptations of invertebrates to terrestrial arctic environments. Det Kongelige Norske Videnskabers Selskab nr. 2:1–44

van't Hoff JH (1894) Études de dynamique chemique. Amsterdam

Honěk A (1996) Geographical variation in thermal requirements for insect development. Eur J Entomol 93:303–312

Honěk A, Kocourek F (1990) Temperature and development time in insects: a general relationship between thermal constants. Zoologische Jahrbücher Systematik 117:401–439

Hopkins GW (1996) Rarity in tree aphids. Ph.D. Thesis, University of East Anglia

Komazaki S (1982) Effects of constant temperatures on population growth of three aphid species, *Toxoptera citricidis* (Kirkaldy), *Aphis citricola* van der Goot and *Aphis gossypii* Glover (Homoptera: Aphididae) on Citrus. Appl Entomol Zool 17:75–81

Komazaki S (1986) The inheritance of egg hatching time of the spirea aphid, *Aphis citricola* van der Goot (Homoptera, Aphididae) on two winter hosts. Kontyû 54:48–53

Kundu R, Dixon AFG (1995) Evolution of complex life cycles. J Anim Ecol 64:245–255

Lamb RJ (1992) Developmental rate of *Acyrthosiphon pisum* (Homoptera, Aphididae) at low temperatures – implications for estimating rate parameters for insects. Environ Entomol 21:10–19

Lamb RJ, MacKay PA, Gerber GH (1987) Are development and growth of pea aphids, *Acyrthosiphon pisum,* in North America adapted to local temperatures. Oecologia 72:170–177

Liu S-S, Hughes RD (1987) The influence of temperature and photoperiod on the development, survival and reproduction of the sowthistle aphid *Hyperomyzus lactucae*. Entomol Exp Appl 43:31–38

Liu S, Meng X (1989) The change patter of development rates under constant and variable temperatures in *Myzys persicae* and *Lipaphis erysimi*. Acta Ecol Sinica 9:182–190

Liu T-X, Yue B (2000) Effects of constant temperatures on development, survival and reproduction of apterous *Lipaphis erysimi* (Homoptera: Aphididae) on cabbage. Southwest Entomol 25:91–99

Mordvilko AK (1908) Beitrag zur Biologie der Pflanzenläuse, Aphididae Passerini. Biol Zentralblatt 28:631–639

Nowieski RM, Gutierrez AP, Yaninck JS (1983) Estimation of thermal thresholds and age-specific life table parameters for the walnut aphid (Homoptera: Aphididae) under field conditions. Environ Entomol 12:680–686

Rakauskas R (2004) Recent changes in aphid (*Hemiptera, Sternorrhyncha:Aphididae*) fauna of Lithuania: an effect of global warming? Ekologija 1:1–4

Reed TD, Semntner PJ (1991) Influence of temperature on population development of two color morphs of the tobacco aphid (Homoptera, Aphididae) on flue-cured tobacco. J Entomol Sci 26:33–38

Rohitha BH, Penman DR (1983) Effect of temperature on the biology of bluegreen lucerne aphid, *Acyrthosiphon kondoi*. N Z J Zool 10:299–308

Summers CG, Coviello RL, Gutierrez AP (1984) Influence of constant temperatures on the development and reproduction of *Acyrthosiphon kondoi* (Homoptera: Aphididae). Environ Entomol 13:236–242

Thieme T, Dixon AFG (2004) The case for *aphis solanella* being a good species. In: Simon JC, Dedryver CA, Rispe C, Hullé M (eds) Aphids in a new millennium. INRA, Paris, pp 189–194

Trudgill DL (1995) Why do tropical poikilothermic organisms tend to have higher threshold temperatures for development than temperate ones? Funct Ecol 9:136–137

Trudgill DL, Honěk A, Li D, van Straalen NM (2005) Thermal time – concepts and utility. Ann Appl Biol 146:1–14

Trudgill DL, Perry JN (1994) Thermal time and ecological strategies – a unifying hypothesis. Ann Appl Biol 125:521–532

Watson SJ (1983) Effects of weather on the numbers of cereal aphids. Ph.D. Thesis, University of East Anglia

Watt AD, Woiwood I (1999) The effect of phenological asynchrony on population dynamics: analysis of fluctuations of British macrolepidoptera. Oikos 87:411–416

Xia J (1997) Biological control of cotton aphid (*Aphis gossypii* Glover) in cotton (inter) cropping systems in China; a simulation study. Ph.D. Thesis, University of Wageningen

Yamaguchi H (1976) Biological studies on the Todo-fir aphid, *Cianara todocola* Inouye, with special reference to population dynamics and morph determination. Bulletin of Government Forest Experimental Station No. 283, Tokyo

de Réaumur RAF (1735) Observation du thérmomètre, faites à Paris pendant l'année 1735, comparées avec celles qui ont été faites sous la ligne, à l'Isle de France, à Alger et en quelquesunes de nos isles de l'Amérique. Mémoires de l'Académie des sciences, Paris 1735:545–576

de Réaumur RAF (1736) Mémoires pour server a l'histoire des insects. 2 vols, Paris

van Straalen NM (1994) Adaptive significance of temperature responses in Collembola. Acta Zool Fenn 195:135–142

Chapter 9
Coccinellids in a Changing World

Helen E. Roy and Michael E.N. Majerus

Abstract Current rates of biodiversity change and loss exceed those historically recorded and show no indication of slowing. Comparisons of recorded extinctions of known species over the last 100 years with rates of extinction of characteristic species in the fossil record indicate that humans have increased the species extinction rate by 100 times. The major drivers of biodiversity change are considered to be habitat change, climate change, invasive species, overexploitation and pollution but changes in biodiversity and ecosystems are most often caused by the interactions between multiple drivers, across both spatial and temporal dimensions. Dietary and habitat generalist and specialist coccinellids should exhibit differing responses to these drivers of change. On continental scales, coccinellids move rather than adapt and when possible, maintain environmental constancy rather than geographic constancy. Only when dispersal is not possible due to barriers of unfavourable habitats must they evolve or die. This means that permanent, relatively continuous latitudinal wildlife corridors are essential to allow species (particularly stenotopic species) to mobilize for survival as environments change. This chapter will review evidence relating specifically to the impacts of drivers of change on coccinellids.

Keywords Coccinellidae · Ladybirds · Environmental change · Habitat destruction and fragmentation · Climate change · Invasive species · Overexploitation · Pollution

9.1 Introduction

> Changes in biodiversity due to human activities were more rapid in the past 50 years than at any time in human history, and the drivers of change that cause biodiversity loss and lead to changes in ecosystem services are either steady, show no evidence of declining over time, or are increasing in intensity. (Millennium Ecosystem Assessment 2005)

Human actions have dramatically transformed virtually all of Earth's ecosystems (Millennium Ecosystem Assessment 2005). Current rates of biodiversity change

H.E. Roy (✉)
NERC Centre for Ecology and Hydrology, Wallingford, Oxfordshire OX10 8BB, UK
e-mail: hele@ceh.ac.uk

P. Kindlmann et al. (eds.), *Aphid Biodiversity under Environmental Change*,
DOI 10.1007/978-90-481-8601-3_9, © Springer Science+Business Media B.V. 2010

and loss exceed by orders of magnitude those historically recorded and show no indication of slowing. Comparisons of recorded extinctions of known species over the last 100 years with rates of extinction of characteristic species in the fossil record indicate that humans have increased the species extinction rate by 100 times (Millennium Ecosystem Assessment 2005). Less direct estimates suggest that current extinction rates are 1,000–10,000 times higher than rates recorded among fossil lineages (Millennium Ecosystem Assessment 2005). Biodiversity change is caused by a range of direct and indirect drivers, whereby direct drivers unambiguously influence ecosystem processes and indirect drivers act diffusely by changing one or more direct drivers. Habitat change, climate change, invasive species, overexploitation and pollution are considered direct drivers with the greatest impact on biodiversity (Millennium Ecosystem Assessment 2005). However, changes in biodiversity and ecosystems are most often caused by the interactions between multiple drivers, across both spatial and temporal dimensions. Following a brief introduction to the biology of ladybirds (Coleoptera: Coccinellidae), this chapter will review evidence relating to the impacts of drivers of change on coccinellids.

9.2 Ladybirds (Coleoptera: Coccinellidae)

Ladybirds belong to the Coleoptera (beetles), a very successful order of insects that evolved about 280 million years ago (Lower Permian). All the modern superfamilies of Coleoptera were established as distinct lineages by the end of the Jurassic. Morphological phylogenies suggest that ladybirds are advanced beetles but molecular evidence suggests that the ladybirds are more closely related to an ancient lineage of Coleoptera, the ground beetles (Carabidae). However, it is widely accepted that the ladybirds constitute a monophyletic family that includes over 5,000 described species worldwide (Howland and Hewitt 1995; Majerus 2003).

There are seven subfamilies of Coccinellidae: Sticholotidinae, Chilocorinae, Scymninae, Coccidulinae, Ortalinae, Coccinellinae, Epilachninae. Most species in the Epilachninae are phytophagous, while the majority of species in the other subfamilies are predatory. The preferred diets of the two feeding stages in the life-cycle, the larval and adult stages, are generally the same. Most predatory ladybirds feed on either aphids or coccids (a few feed on both), however some predatory species feed on mites, adelgids, aleyrodids, ants, chrysomelid larvae, cicadellids, pentatomids, phylloxera, mycophagous coccinellids and psyllids (Dixon 2000). One wingless and blind species, *Cleidostethus meliponae*, has been found in the nests of the bee, *Melipona alinderi*, in East Africa (Salt 1920). A small number of species within the Coccinellinae and Epilachninae are mycophagous, feeding on the hyphae and spores of fungi. There is also considerable variability in the degree of dietary specialisation between species. Thus, whereas some species have a very narrow range of preferred prey, such as a single species of mite, aphids of a single genus, or plants of a single family, other species have a wide prey range. For example, *Harmonia axyridis* will feed on aphids, coccids, adelgids, psyllids, many other insects, including other coccinellids and lepidopterans, and also mites. Adaptations to specific or more general diets are complex, including mandibular dentition, gut length and structure, and morphological features that affect mobility. When preferred prey are scarce, many predatory coccinellids will feed on alternative foods, such as pollen, nectar, honey-dew and fungi, and many also resort to cannibalism (Hodek 1996a).

There is a strong interaction between the dietary preferences of coccinellids and their geographic distributions. At a global scale, for example, aphids are more abundant than coccids in temperate regions and a higher proportion of ladybirds in this climatic zone specialize on aphids than coccids (Dixon 2000). The reverse is the case in the wet tropics. Locally, ladybird densities are strongly correlated to food availability (Majerus 1994).

Ladybirds are holometabolous insects and as such begin their life cycle as an egg, which hatches to give rise to a larva that generally passes through four instars before pupation and metamorphosis into an adult. With only a few exceptions the aphidophagous species lay their eggs in clusters and the coccidophagous species lay their eggs singly or in very small groups. Development rate is dependent on temperature and food quality (Majerus 1994; Dixon 2000). Although the relative lengths of time that coccidophagous and aphidophagous coccinellids spend as egg, larva, pupa and adult are similar for a given temperature, ladybirds specializing on aphids develop faster than coccidophagous, phytophagous or mycophagous species (Dixon 2000).

Coccinellids are found on all the continents of the world with the exception of Antarctica. In the context of this chapter, the distribution of ladybirds needs to be considered at both geographic and local scales. The current geographical distributions of ladybird species are the result of historical events relating to climatic changes, vegetation changes, geographic topography, plate tectonics, and more recently, anthropogenic impacts on land usage (Majerus 1994). Locally, coccinellid distributions depend on food availability, habitat preferences and the degree of habitat specialisation. Thus, while some species are habitat generalists (eurytopic), and may be found in a diverse array of environments, others are restricted to one or a small number of habitat types (stenotopic), and display a suite of adaptations to these specific habitats. *Myzia oblongoguttata* and *Myrrha octodecimpunctata* only breed on needled conifers and, therefore, are stenoptopic. In contrast, *H. axyridis* is little restricted by habitat and so is eurytopic. The distributions of some coccinellids are further affected by the distributions of non-prey species. Thus, for example, some species have little or no tolerance to aggressive ants, and are excluded from foraging within the ranges of these ants. Conversely, other coccinellids, such as *Coccinella magnifica* and *Platynaspis luteorubra* are myrmecophilous, having specific adaptations for living in close proximity to ants (Sloggett and Majerus 2003; Völkl 1995).

Most coccinellids exist in complex ecosytems. Aspects of their life-histories, dietary preferences and habitat preferences, as well as their behaviour, all have pertinence to the issue of coccinellids in a changing world. For example coccinellids vary in mobility and habitat plasticity, particularly the ways in which they cope with unfavorable habitats (*sensu* Southwood 1977), and their interactions with their enemies and competitors and, consequently, will differ in their responses to global drivers of change.

9.3 Main Drivers of Biodiversity Change

9.3.1 Habitat Destruction and Fragmentation

Nearly all of Earth's ecosystems have been dramatically altered through human actions (Millenium Ecosystem Assessment 2005). Indeed the Millennium Ecosystem Assessment (2005) highlights some striking statistics: more land was converted to cropland in the 30 years after 1950 than in the preceding 150 years; 35% of mangrove swamps have been lost in the last two decades; over half of the 14 biomes assessed have experienced a 20–50% conversion to human use, with temperate and Mediterranean forests and temperate grasslands being the most affected. The rate of destruction of some ecosystems has begun to slow but in some cases this is because little habitat remains. Moreover, remaining areas of habitat are often fragmented. Larger remnants and fragments that are close to other remnants exhibit less biodiversity loss than small and unconnected fragments, which are more vulnerable to species extinctions. Furthermore, the edges of habitat fragments, have different climates and favour different species compared to the habitat interior; small fragments are unsuitable for species that favour the habitat interior (Woodroffe and Ginsberg 1998; Laurance et al. 2001). Stenotopic species, which exhibit a narrow relationship with a particular habitat, could be at greater risk from habitat destruction than eurytopic species, particularly if their dispersal ability is also weak. However, the consequences of fragmentation and community dynamics in response to fragmentation are difficult to assess because of the complexity caused by biotic and abiotic interactions at multiple scales (McGarigal and Cushman 2002).

Habitat fragmentation is known to affect arthropods differently depending on the trophic guild with which they are associated and the degree to which they specialize (Davies et al. 2001). Species with a wide host range are more likely to find sufficient resources within a fragment compared to specialists. In addition, generalists are more likely to utilize corridors connecting fragments for nutrition and harborage than are specialists. The habitat and food requirements of different species of coccinellid are diverse and many species are known to be restricted to specific habitats (Honěk and Hodek 1996). At a geographic scale, for example, *Coccinella alta* and *Bothrocalvia lewisi* are high montane species, while *Hippodamia arctica* and *Anisosticta strigata* are sub-arctic species. In Europe, *Anisosticta novemdecimpunctata* and *Macronaemia episcopalis* are restricted to the vegetation of wetlands (fen, marsh and moist meadows); *Adalia conglomerata* and *M. oblogoguttata* to coniferous forests; *Coccinella magnifica* to habitats close to *Formica* ant nests (Honěk and Hodek 1996; Majerus et al. 2006a). A small proportion of coccinellids are much more eurytopic, with species such as *Adalia bipunctata* and *H. axyridis* occurring in a wide range of arbivorous and lower herbage habitats, including many crop systems. The proportion of specialist to generalist species has not been assessed globally, but in Britain, of 42 species, Majerus (1994) categorised 11 to be habitat generalists, ten to be generalists within the constraint of breeding either on trees or on low growing herbage, but not both, and 21 to be habitat specialists.

Studies based on metapopulation and island biogeography theory predict that species occupying higher trophic levels will be more susceptible to the effects of habitat loss than those occupying lower levels (Dobson et al. 2006). It has been demonstrated that parasitoids tend to be more extinction-prone than their herbivore hosts (Cronin 2004) and declines in diversity and abundance of parasitoids result in declines in parasitism rates with increasing fragment isolation or fragment size (Rand and Tscharntke 2007). The loss of higher trophic levels ultimately disrupts top down control of prey populations in natural habitat fragments (Dobson et al. 2006). Many species of ladybird are predatory, occupying higher trophic levels, and have been shown to respond directly to fragment isolation in experiments (Kareiva 1987; Banks 1999). Ladybirds are further predicted to respond to landscape level factors such as habitat fragment area, shape and the structure of neighbouring fragments (Stoner and Joern 2004). In a study on insect communities in tall grass prairie remnants, Stoner and Joern (2004) demonstrated that fragment shape was an important variable in explaining changes in the community composition of coccinellids. This study concluded that species associated with higher trophic levels, such as ladybirds, are more susceptible to fragmentation because of the challenges of recolonisation following local extinction events.

Conversely, it has been speculated that predictions based on island or metapopulation based theory may not apply to generalist predators because they can exploit the intervening landscape matrix (Rand et al. 2006). Indeed, in situations where anthropogenically modified matrix habitats provide resources for generalist predators, such as agricultural systems, spillover into adjacent natural systems could result in increased predator abundance (Rand et al. 2006). Abundance of *Coccinella septempunctata* and *Propylea quatuordecimpunctata*, two relatively generalist ladybird predators, was shown to be decoupled from local resource density but was consistent with spillover effects (Rand and Tscharntke 2007). Both species were more abundant within nettle patches in simple compared to complex habitats, suggesting that these species benefit from the surrounding agricultural habitats (Rand and Tscharntke 2007).

It can be concluded that dietary and habitat generalist and specialist coccinellids should exhibit differing responses to habitat loss and fragmentation. It is widely accepted that some species of coccinellid are dominant predators, impacting aphids in natural habitats as a consequence of spillover from cropland habitats (Kareiva 1987; Müller and Godfray 1997). In addition, coccinellids are highly voracious compared to many other natural enemy taxa (Müller and Godfray 1997) and so their impact on prey populations is likely to be disproportionate to their abundance (Rand and Tscharntke 2007). However, there is an absence of empirical work examining the effects of habitat loss on the abundance of a broad assemblage of natural enemies, including coccinellids. Understanding the cumulative effects of habitat loss on predator-prey interactions, and even single species, requires such an approach. Meanwhile the literature is in conflict. The prevailing view that the loss of natural habitat will disrupt predator-prey interactions in natural systems is disputed by Rand

and Tscharntke (2007). This ambiguous view is undoubtedly a consequence of the concentration of research on a relatively small number of coccinellid species that have economic importance as biological control agents. For example, in an analysis of the literature on coccinellids published between 1995 and 2004, Sloggett (2005) showed that in 623 papers found, only 62 species were studied. Moreover 40% of the studies described concerned just two species, *C. septempunctata* and *H. axyridis*, with three further species, *Coleomegilla maculata*, *A. bipunctata* and *Hippodamia convergens* together comprising another 29% of the studies. All of these species are eurytopic and aphidophagous. Given this extreme knowledge bias, it is obvious that research is needed on the habitat requirements and responses to habitat change of a suite of coccinellids with differing levels of habitat specialization before informed conclusions on the effect of habitat destruction and fragmentation can be made. Undoubtedly, the impact of habitat destruction and fragmentation on a particular species of coccinellid will crucially depend on the degree of specialization of that species.

9.3.2 Climate Change

Recent changes in climate, particularly warmer temperatures, have already begun to impact on biodiversity and ecosystems. Changes in species distributions, population sizes, the timing of reproduction and migration events, and in the frequency of pest and disease outbreaks have all been documented and linked to elevated temperatures. The Intergovernmental Panel on Climate Change predicts an increase in global mean surface temperature of 2–6.4°C above pre-industrial levels by 2100, increased incidence of floods and droughts and a rise in sea level of an additional 8–88 cm between 1990 and 2100. In some regions, ecosystem services could be enhanced by low levels of climate change but at high levels detrimental impacts will vastly outweigh the benefits in most biomes (Millenium Ecosystem Assessment 2005). Indeed, climate change scenarios are predicted, with medium to high certainty, to increase the loss of biodiversity and the risk of extinction for many species (Parmesan and Yohe 2003). There are a number of factors that will exacerbate the situation for some species: low population sizes, limited range of accepted food, high degree of habitat or host plant specificity and limited climatic range.

The importance of the relationship between ladybirds and temperature is widely recognized. Interestingly, the names adopted by some nations for these much loved insects exemplify this; in the Czech Republic they are called slunéčka (small suns) and in Japan they are known as tento mushi (sun loving insects). The importance of climate, particularly microclimate, to coccinellid development and behaviour has been demonstrated in a number of studies. Smith (1971) working in maize, showed that the temperature difference between low density stands and cooler high density stands could be as much as 8.6°C. *Coccinella transversoguttata* and *Coccinella novemnotata* preferred the higher temperature, low density stands, while *C. maculata* and *Hippodamia tredecimpunctata* sheltered from the hottest temperatures in the high density crop. Honěk (1979) found a similar situation in temperate cereal crops, where *C. septempunctata* and *Coccinella quinquepunctata* preferred sparse and well-insolated stands with a warm microclimate, while *P. quatuordecimpunctata* tolerated dense stands with a much cooler microclimate. Ewert and Chiang (1966) demonstrated that ladybirds also exhibit humidity preferences. The vertical distribution of *H. convergens*, *H. tredecimpunctata* and *C. maculata* in barley corresponds to preference for specific humidities within the crop. The humidity preferences of these species were correlated with their resistance to desiccation. The daily movement of ladybirds on host plants further illustrates the importance of microclimate. For example, Benton and Crump (1981) observed a regular pattern of *C. maculata* climbing up the vegetation in the morning, and returning to lower levels in the afternoon. Some species, such as *C. septempunctata*, increase their body temperature through thermoregulatory behaviour that leads them to seek out bare ground, or insolated regions of the host plant (Honěk 1985).

Oviposition behaviour also reflects microclimatic preferences. *Coccinella septempunctata* often lay eggs on bare, insolated soil or upper leaf surfaces (Ferran et al. 1989). *Adalia bipunctata* and *Anatis ocellata* lay eggs on lower leaf surfaces, which is thought to be an anti-predator strategy, but the lower leaf surface also provides a humid and cool environment (Hodek 1996a). Risk of predation on the lower leaf surface is likely to be lower than on the upper leaf surface because predatory insects will forage primarily on the warmer, upper leaf surface. The reduced risk of mortality may outweigh the advantage of faster development that could be achieved on the upper leaf surface. Indeed, Lövei et al. (1991) demonstrated that ladybirds select more sheltered oviposition sites as temperature increases. Therefore, it appears that at elevated temperatures the cost-benefit balance tends towards oviposition site selection that reduces predation at the expense of rapid development.

Coccinellids usually survive seasonally unfavourable conditions in the adult stage. For ladybirds, unfavourable conditions may be a consequence of lack of food or adverse climatic factors, i.e. low temperatures, very high temperatures, or very dry conditions. If climatic conditions are favourable when food becomes scarce, ladybirds generally disperse to seek food elsewhere. Conversely, under unfavourable climatic conditions, ladybirds become inactive, some species entering programmed diapause, while other species simply become dormant or quiescent. The difference between these states is important in the context of potential climatic change. Quiescence involves a response to periods of sudden, unpredictable weather; the ladybirds simply become inactive but resume activity as soon as conditions become favourable again. Dormancy (also known as oligopause) is a response to unfavourable conditions that are seasonally predictable. Ladybirds become inactive, generally for a fixed period, surviving on fat reserves, but they are capable of becoming active to feed for short periods within the unfavourable period if conditions permit. Diapause is also a response to predictably unfavourable seasons, and follows a preparatory period in which ladybirds build up their fat reserves and either do not mature their gonads, or reabsorb their eggs. Ladybirds in diapause do not become active in brief favourable periods, and cannot reproduce until diapause is terminated.

In some species, a diapause before reproduction is obligatory and genetically programmed (Dobrzhanskii and Dobzhansky 1922), but for many species, induction of diapause is a very plastic response initiated by one or more environmental factors. The factors that induce dormancy or diapause vary among species of coccinellid, with the nature and availability of food, day length, temperature, humidity and the physiological state of host plants all being involved in some species (reviewed by Hodek 1996b). The likelihood that an individual ladybird survives through a dormant period is dependent on the level of reserves that it has accumulated prior to dormancy (Hodek and Cerkasov 1963; Barron and Wilson 1998). Fat, glycogen and water reserves all diminish during dormancy, with rates of reduction being affected by changes in ambient temperature, and in the case of water loss, humidity. For overwintering coccinellids, fluctuating temperatures or abnormally high temperatures generally cause reserves to be used up faster than do constant low temperatures (Majerus 1994). Consequently, increases in winter temperature or in fluctuations in winter temperatures due to climate change may lead to increasing rates of winter mortality in species with genetically controlled obligatory diapause.

Some coccinellids make lengthy migration flights to and from dormancy sites, but others remain in, or close to, their preferred breeding habitats during unfavourable seasons. Some of the species that migrate before becoming dormant form large overwintering (e.g., *H. convergens*) or aestivating (e.g. *Micraspis frenata*) aggregations. Such behaviour is most common in species that feed on aphids, which are notoriously ephemeral. These overwintering sites are often at high altitude, where temperatures through the winter are low and so the rate of depletion of reserves is minimised. In some species, such as *H. convergens* in California, movement to and from montane overwintering sites is dependent upon specific autumn and spring winds (Hagen 1962). Such movement may be perturbed by climate changes to the detriment of the ladybird populations.

The relationship between coccinellids and climatic factors are complex, varying temporally and spatially, and with life stage. Although it may be expected that ladybirds will be negatively affected by climate change, direct empirical evidence is sparse. However, at a local level there are some indications that climatic changes are already influencing annual cycles. In Britain, survey data from 1980 to the present day suggest that *C. septempunctata* movement from overwintering sites, earliest mating and earliest eclosion have all shifted 11–18 days earlier in spring over the last 25 years (Majerus unpublished data). There is no indication whether these changes have had a positive or negative effect on populations of this species. However, such rapid changes do have the potential to temporally decouple predators from peaks in abundance of their preferred prey. There are several potential consequences of such a decoupling: aphid populations may suffer reduced predation by coccinellids, intraguild predation may increase as coccinellids turn to alternative prey, and cannibalism may increase, reducing coccinellid numbers.

Some of the more generalist species of ladybird have been shown to exhibit phenotypic plasticity, which could be adaptive under scenarios of climate change, and help buffer these species against some of the detrimental effects of changing climatic factors. *Adalia bipunctata* is a highly polymorphic species (Majerus 1994), in which the various genetic forms can be broadly classified into melanics (black with red spots) or non-melanics (red with black spots). The frequency with which these colour morphs occur varies geographically (Brakefield 1984). Clines in melanism in The Netherlands have become shallower (relative decrease in melanic forms) over time (1978–1998) and this change has coincided with an increase in local ambient spring temperatures (De Jong and Brakefield 1998). The authors invoke the thermal melanism hypothesis, whereby an increase in temperature is predicted to reduce the contribution of radiation to warming up of ladybirds. It is speculated that lower minimum temperatures (east Netherlands) are consistent with a large advantage to melanics when warming up early in the day. Other highly polymorphic species include *H. axyridis*, *Cheilomenes sexmaculatus*, and *Propylea japonica*, all of which are eurytopic.

Some species are euryotopic near the centre of their geographic distributions and stenotypic at the edges of their range (Majerus 1994). Indeed, close to the edge of their range they may only be able to survive through adaptation to some specific type of habitat. *Coccinella quinquepunctata* appears highly eurytopic in continental Europe, but in Britain (Fig. 9.1) its known distribution is confined to unstable river shingles in west Wales and the Spey Valley in Scotland. Global warming could result in *C. quinquepunctata* increasing in distribution across Britain. *Hippodamia tredecimpunctata* (Fig. 9.2) is sporadically reported on the south and east coasts of Britain where it appears to arrive from continental Europe and establish small colonies that soon become extinct (Majerus 1994). Climate change may be favourable for the long term establishment of *H. tredecimpunctata* in England in the future. *Halyzia sedecimguttata* has increased in abundance in Britain over recent years. This mildew feeding species was considered to primarily feed on mildews on sycamore but recently has been found more widely on deciduous trees (UK Ladybird Survey, unpublished data). Warmer, wetter springs in Britain could have increased the prevalence of mildew and the distribution of *H. sedecimguttata* could have increased in response to this available food. The UK Ladybird Survey (www.ladybird-survey.org) will be investigating this further. There will undoubtedly be winners and losers as a consequence of climate change but the exact nature and balance of outcomes is hard to predict because of the complexity of interactions and lack of available data.

Climatic factors are likely to influence the dynamics of interactions between ladybirds and their prey, predators and parasites. For example, short range clinal variation in the prevalence of a sexually transmitted Laboulbeniales fungus (*Hesperomyces virescens*) infecting *A. bipunctata* is associated with urbanization and has been linked with increased temperatures within the urban environment (Welch et al. 2001). A second sexually transmitted parasite of *A. bipunctata* may also be affected by climate change. The ectoparasitic mite *Coccipolippus hippodamiae* infects *A. bipunctata* over much of continental Europe, but is absent from Britain (Webberley et al. 2006). Hurst et al. (1995) have attributed the absence of this mite from British populations of *A. bipunctata* to the lack of transmission between generations in the occasional years when the species is fully univoltine (in most of continental Europe, *A. bipunctata* has two or more overlapping generations per year) and the old generation dies before the new generation is reproductively

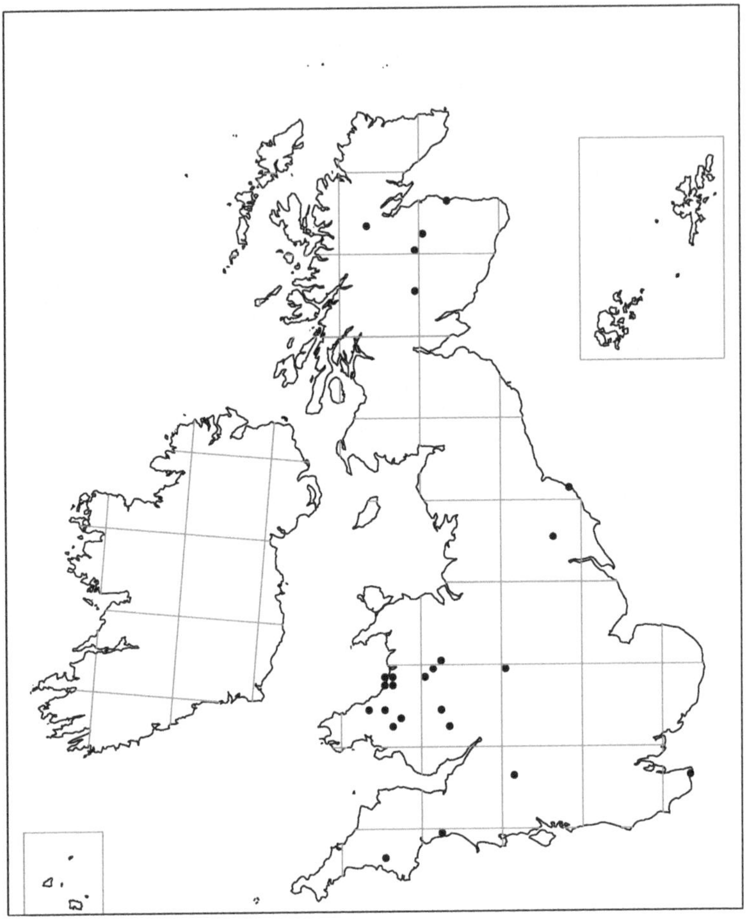

Fig. 9.1 Distribution of *Coccinella quinquepunctata* in Britain where it is confined to unstable river shingles in west Wales and the Spey Valley in Scotland (data from 1832 to present). Data from the UK Ladybird Survey, Biological Records Centre

active. From surveys over the last 28 years, such years are becoming less common, at least in southern England (Majerus unpublished data). Increasing temperatures are, therefore, likely to increase the probability of long-term survival of this mite in British populations of *A. bipunctata*. As *C. hippodamiae* has a severe impact on the female ladybirds it infects – eggs laid by females more than 3 weeks after becoming infected fail to hatch (Hurst et al. 1995) – this is likely to have a negative impact on *A. bipunctata* populations.

Studies of the effects of current climate changes on coccinellids at a more global scale are lacking. However, examination of coccinellid remnants in Quaternary deposits allows for some inferences. The Quaternary Period, from 2.4 million years ago to present, has been characterised by numerous intense climatic oscillations, particularly at higher latitudes. Climatic oscillations resulted in many coccinellids

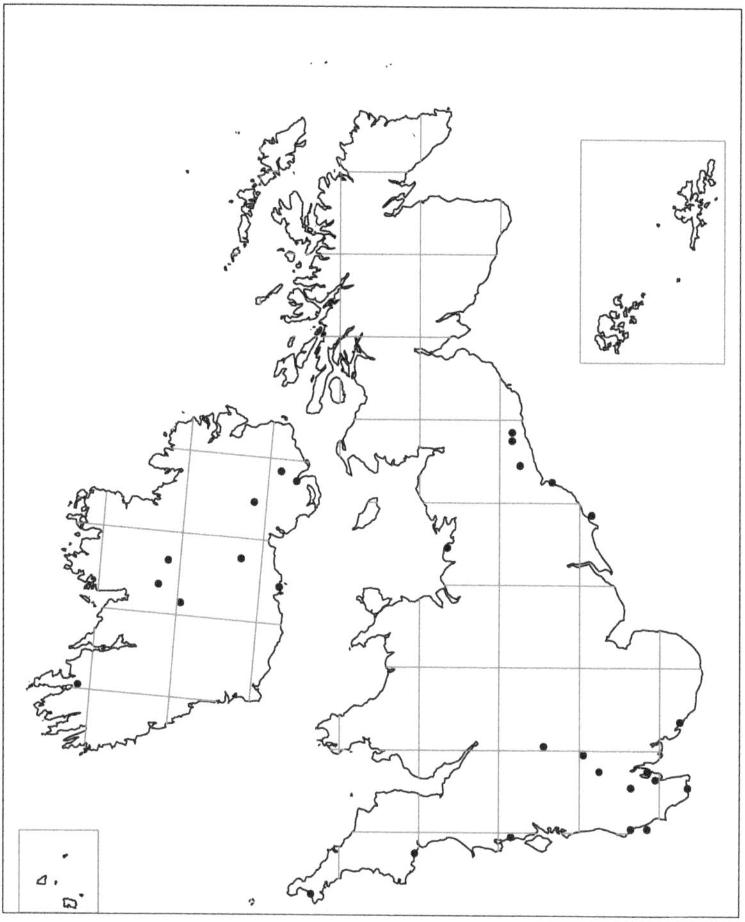

Fig. 9.2 Distribution of *Hippodamia tredecimpunctata* in Britain where it appears to arrive from continental Europe, establishes small colonies and these then become extinct (data from 1819 to present). Data from the UK Ladybird Survey, Biological Records Centre

changing their geographic ranges dramatically, particularly with respect to latitude, but also in some cases with respect to altitude. For example, *H. arctica* was common in Britain 40,000 years ago during the Upton Warren Interstadial and up until about 25,000 years ago, when it disappeared as temperatures declined further during the Late Weichselian Glaciation. A little over 13,000 years ago it reappeared, presumably from the south, moved northwards through Britain as temperatures increased, and disappeared again, only to reappear from 11,000 to 10,000 years ago in the cooler Younger Dryas period (Osborne 1971). It then disappeared yet again, and has not reappeared. In Europe, the species is now restricted to the north of Sweden and Norway, mainly within the Arctic Circle. A similar pattern is seen in *A. strigata* (Majerus 1994). The enormity of potential geographic range alterations is illustrated by the ladybird *Ceratomegilla ulkie* which was recorded from an organic silt from the River Thames around 40,000 years ago (Briggs et al. 1985), is now absent from Europe, but still occurs at high latitudes or high altitudes in north-western Canada and north-eastern Asia.

This pattern suggests that, on continental scales, coccinellids move rather than adapt. When possible, they maintain environmental constancy rather than geographic constancy. Only when dispersal is not possible due to barriers of unfavourable habitats must they evolve or die. This means that permanent, relatively continuous latitudinal wildlife corridors are essential to allow species (particularly stenotopic species) to mobilize for survival as climate changes. The alternative is that these species will have to adapt to new environmental conditions, and due to the speed of the climatic changes that we are now experiencing, it is likely that some stenotopic species will simply not be able to adapt quickly enough to avoid extinction.

9.3.3 Invasive Species

> I have an unnerving nagging at the back of my mind that faunal and floral homogenisation is among the most detrimental of the ways in which humanity is affecting the Earth (Majerus 1994, p 311)

The distribution of species on Earth is becoming increasingly homogeneous for two main reasons: species are experiencing higher rates of extinction and rates of invasion and introduction are accelerating with increased trade and transportation (Millenium Ecosystem Assessment 2005). Invasion and extinction are not mutually exclusive; invasive alien species have been a major cause of extinction, particularly on islands and in freshwater habitats. Many coccinellids are threatened by invasive species and a small number of species have been implicated as invasive species themselves.

Classical biological control involves the introduction of a non-native natural enemy to control an exotic pest. Ladybirds have a long history of use as biological control agents against pest insects (Majerus 1994). In 1888, the Australian vedalia ladybird, *Rodolia cardinalis*, was released in California to control cushiony scale, *Icerya purchasi*, which were having devastating impacts on the citrus industry (DeBach 1964; Majerus 1994). This ladybird established and effectively reduced the scale insect population, an event widely regarded as marking the advent of modern biological control. Indeed, ladybirds are often considered flagships of biological control and the predatory habits of many species of ladybird have no doubt contributed to their popularity. Ladybirds are much-loved worldwide, but there is one species *H. axyridis*, that has become extremely unpopular outside of its native range over the last two decades for a number of reasons (Roy et al. 2006; Majerus et al. 2006b). First, it threatens the diversity of native aphidophagous species through direct competition for prey and as a voracious, intraguild predator (Majerus et al. 2006b; Pell et al. 2008). Second, it is considered a pest species in the USA because at times it feeds on soft fruits, adversely affecting the wine and fruit growing industries (Koch 2003; Koch et al. 2004; Kovach 2004). Third, it overwinters at high densities in buildings (thousands of individuals) and is consequently regarded as an urban nuisance (Kovach 2004). In addition, it is likely to adversely impact many non-target aphids and coccids, and has been shown to attack the immature stages of butterflies and many other insects. *Harmonia axyridis* has become perhaps the most infamous of invasive insects in the twenty-first century.

Harmonia axyridis was first introduced into the USA as a biological control agent of pest insects in 1916 (Gordon 1985). Many further releases were made, but the species failed to establish until the 1980s, when it was found in south-eastern Louisiana (Chapin and Brou 1991). It is now widespread across the USA and has spread into Canada (Koch and Galvan 2008). Intentional introductions of *H. axyridis* were made across Europe (Belgium, Czech Republic, France, Germany, Greece, Italy, Netherlands, Portugal, Spain and Switzerland) beginning in the early 1980s and the species is now considered established in all these countries, with the exception of Portugal and Spain (Brown et al. 2008). Inevitably, *H. axyridis* arrived in countries where it had not been deliberately introduced. By 2004. it had arrived in England and Luxembourg; in 2006 it was reported in Austria, Denmark, Wales and Norway; in 2007 Liechtenstein, Sweden and Northern Ireland (Brown et al. 2008). It has also been introduced and established in Egypt, South Africa and Argentina. In many parts of North America and Western Europe it is now the most abundant coccinellid species (Fig. 9.3).

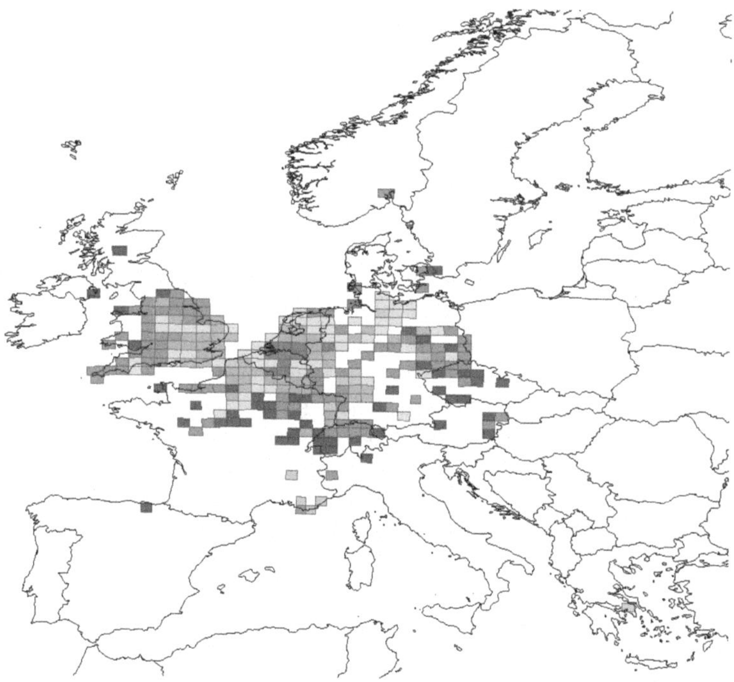

Fig. 9.3 European distribution of *Harmonia axyridis*. Adapted from Brown et al. (2008)

The success of *H. axyridis* as an invasive species is attributed to a number of factors including its size, voracity, reproductive potential, dispersive ability, eurytopy, and polyphagous habits (Evans 2000; Majerus et al. 2006b). Moreover, *H. axyridis* is a unidirectional intra-guild predator (Hironori and Katsuhiro 1997; Cottrell and Yeargan 1998; Phoofolo and Obrycki 1998; Michaud 2002a; Sato et al. 2003; Felix and Soares 2004; Snyder et al. 2004; Amkar 2006; Kajita et al. 2006; Cottrell 2007). As such, it is thought to be an important force in structuring aphidophagous guilds through direct impacts on other species of coccinellid (Yasuda and Shinya 1997; Kajita et al. 2000) and their prey (Majerus et al. 2006b).

Ware and Majerus (2008) examined intra-guild predation (IGP) of immature stages of British and Japanese coccinellids by *H. axyridis*. Thirteen species of coccinellid were included in this study: *H. axyridis*, *C. septempunctata*, *A. bipunctata*, *Adalia decempunctata*, *P. quatuorodecimpunctata*, *Calvia quatuordecimguttata*, *A. ocellata*, *Harmonia quadripunctata*, *C. quinquepunctata*, *C. septempunctata brucki*, *C. sexmaculatus*, *P. japonica* and *Eocaria muiri*. First instar *H. axyridis* larvae were considered unidirectional intra-guild predators of first instar *A. decempunctata*, *P. quatuordecimpunctata* and *C. quinquepunctata* and bidirectional intra-guild predators of *C. septempunctata*, *A. bipunctata*, *C. quatuordecimguttata* and *H. quadripunctata*. The only unidirectional intra-guild predator of *H. axyridis* was *A. ocellata*, a large British coccinellid (Majerus et al. 2006b). Fourth instar *H. axyridis* dominated as unidirectional intra-guild predators of other fourth instar coccinellid larvae (Ware and Majerus 2008).

These studies indicate the potential for *H. axyridis* to disrupt the community structure of aphidophagous coccinellids. Furthermore, the complexity and unpredictability of ecological impacts of invasive intraguild predators contribute to a high risk of adverse interactions when such species are intentionally introduced (Snyder and Evans 2006). *Harmonia axyridis* has been shown to be a dominant intra-guild predator in the field and in its native Japan it has limited the establishment of the introduced *A. bipunctata* (Sakuratani et al. 2000; Kajita et al. 2006). In contrast, *H. axyridis* has become abundant and widely distributed through North America and Europe (Brown et al. 2008) and is negatively impacting native ladybird species (LaMana and Miller 1996; Elliott et al. 1996; Brown and Miller 1998; Colunga-Garcia and Gage 1998; Michaud 2002a; Nault and Kennedy 2003). *Coccinella septempunctata* has also become established widely in North America but *H. axyridis* is considered to be a more aggressive predator than *C. septempunctata* (Yasuda et al. 2004). Indeed, larvae of *H. axyridis* consume those of *C. septempunctata* but the reverse is considerably less frequent (Hironori and Katsuhiro 1997). Consequently, *H. axyridis* represents a considerable threat to indigenous ladybirds through intra-guild predation (Snyder et al. 2004).

9.3.4 Overexploitation

The one direct driver that is unlikely to have any strong negative impacts on coccinellids is overexploitation. This is because the coccinellids that have been exploited for biological control have a high reproductive potential, and because releases usually involve large numbers of ladybirds that have been cultured in captivity from relatively small samples of field-collected material. The one exception to this involves augmentative biological control using the North American coccinellid, *H. convergens*. This species has been collected in huge numbers from overwintering aggregations in mountains in America for over a 100 years. The ladybirds collected are bottled alive and stored in carefully controlled conditions until the spring when they are sold, by the pint or gallon, through garden centres or by mail. This practice does not appear to have had any detrimental effects on populations of *H. convergens* although the ladybirds are translocated away from their native habitat and released to control aphid species that may not even comprise suitable prey. For example, in the 1990s, a shipment of 30 million *H. convergens* were released onto *Tilia* trees lining roads in Dutch towns, to reduce aphid populations and thereby reduce amounts of aphid honeydew that fell onto parked cars. The project failed because the ladybirds all disappeared within a few weeks. One has to question the sense of releasing a species that relies on montane overwintering sites into a country that is devoid of mountains.

Although ladybirds have not suffered from overexploitation, it could be argued that the ill-advised releases of inappropriate species, such as *H. axyridis*, comprise cases of, if not overexploitation, then very unwise exploitation.

9.3.5 Pollution

Over the last 40 years pollution, particularly nutrient loading into ecosystems, has emerged as one of the most important drivers of change (Millenium Ecosystem Assessment 2005). More than half of all the synthetic nitrogen fertilisers ever used on Earth have been used since 1985 (Millenium Ecosystem Assessment 2005). Nitrogen application has increased five-fold since 1960 and up to 50% of the nitrogen applied will be lost to the environment and contribute to nitrous oxide emissions, eutrophication and hypoxia in marine environments (Millenium Ecosystem Assessment 2005). Global atmospheric CO_2 has risen from 280 μl/l in 1,700 to 370 on μl/l (Chen et al. 2007). It is predicted that elevated CO_2 will have profound impacts on species and communities altering trophic interactions between plants, herbivores and natural enemies. There have been many studies on the responses of plants and herbivores to elevated CO_2 (Coviella and Trumble 1999) but studies on higher trophic levels are lacking. Chen et al. (2007) assessed the impact of elevated CO_2 on *H. axyridis* and found no effect on growth and development. However, Chen et al. (2005) found that *H. axyridis* larvae consumed more cotton aphids, *Aphis gossypii*, fed elevated CO_2-grown cotton plants in order to compensate for the reduced soluble protein in cotton aphid. Plants grown in elevated CO_2 treatments have decreased foliar N and an increased C:N ratio. In contrast, elevated CO_2 did not influence aphid consumption rate by *H. convergens* (Stacey and Fellowes 2002). Clearly, there is a need for further studies on more species and with a greater range of environmental pollutants before any generalisations can be made.

The use of insecticides with low toxicity to natural enemies, including predatory coccinellids, is an important component of integrated pest management (Grafton-Cardwell and Gu 2003; Galvan et al. 2005). Consequently a number of studies have evaluated the toxicity of insecticides to species of coccinellid that are deemed agriculturally important by providing ecosystem services, namely pest control. *Harmonia axyridis* exhibits tolerance to fungicides (Michaud 2001), acaricides (Michaud 2002b) and some insecticides (Michaud 2002c). However, densities of *H. axyridis* were higher in field plots treated with spinosad or indoxacarb (new insecticides with high toxicity to pest insects and negligible toxicity to natural enemies) than in those treated with conventional insecticides such as carbaryl, chlorpyrifos, bifenthrin and λ-cyhalothrin (Galvan et al. 2005). Indeed, conventional chemicals were highly toxic to *H. axyridis*. The least toxic chemical insecticide to *H. axyridis* is spinosad; higher densities of *H. axyridis* were found in the spinosad-treated plots than the indoxacarb plots (Ochou and Martin 2003; Galvan et al. 2005). Toxicity of insecticides to this coccinellid varies with life stage: eggs and pupae are the most resilient stages to chemical insecticides and larvae followed by adults, which are the most susceptible (Youn et al. 2003).

Most studies on non-target impacts of insecticides consider only direct mortality through acute toxicity. Further work is required to examine residual activity and potential sublethal effects (such as reduced fecundity or disruption of predatory behaviour). Adult *Rodolia cardinalis* exposed to neonicotinoid imidacloprid and the pyrethroid cyfluthrin exhibited high mortality and progeny production was reduced by these chemicals and also fenpropathrin and buprofezin (Grafton-Cardwell and Gu 2003). Larval development to adulthood was also reduced after exposure to all these chemicals and also to pyriproxifen. Interestingly, neem-fed prey were not toxic to *Harmonia conformis* but were to the neuropteran *Mallada signatus* (Qi et al. 2001) indicating the specific nature of insecticidal activity.

In the UK, 47% of the 1.12 million tons (dry solids) of sewage sludge produced annually is applied to agricultural land as a fertiliser (Gendebien et al. 1999; Green et al. 2003). However, sludge contains a wide range of potentially toxic elements including cadmium and zinc. Indeed, although trace metals enter agroecosystems from other sources (atmospheric deposition and use of agrochemicals), sewage sludge is the principal source of trace metal addition to agricultural soils (Nriagu and Pacyna 1988). Green et al. (2003) examined the transfer of cadmium and zinc from sewage sludge amended soil through a plant-aphid system to newly emerged adult *C. septempunctata*. Cadmium was not biomagnified by the crop during transfer from soil to aphid but zinc was, resulting in concentrations four times greater in the aphids. However, there was no increase in the body burden of either metal in newly emerged *C. septempunctata*, larvae of which sequestered zinc in the exuviae and consequently regulated the body burden of zinc effectively (Green et al. 2003). The authors conclude that from an economic and ecological perspective soil trace metal concentrations added through sewage sludge application remain below levels detrimental to coccinellids. However, Green et al. (2003) add a precautionary note that is applicable to many studies relating to pollution effects on coccinellids: it has been demonstrated that increased trace metal concentrations in the diet of coleopteran larvae can cause changes in adult locomotive behaviour without significantly affecting the body burden of the adult. This highlights the need to assess the subtle interactions (on a species by species basis) that may be at play, but are often overlooked, in these systems. Indeed a study on copper-based fungicides (used for controlling foliar diseases in citrus production) demonstrated that three species of coccinellid (*Curinus coeruleus*, *H. axyridis* and *Olla v-nigrum*) exposed to residues survived to adulthood, but larvae of *O. v-nigrum* experienced a significant increase in developmental time, pre-reproductive period and decrease in fecundity than unexposed beetles. The other two species were unaffected by exposure (Michaud and Grant 2003).

9.4 Summary

Biodiversity and ecosystem services experience change due to natural causes, but current changes are dominated by anthropogenic drivers. The most important direct drivers of biodiversity loss and ecosystem service changes are habitat change, climate change, invasive alien species, overexploitation and pollution (Millennium Ecosystem Assessment 2005). All these, with the exception of overexploitation, impact coccinellid populations. However, empirical evidence is sparse or lacking for most of these drivers in respect of the vast majority of coccinellid species. Consequently, it is not possible to reliably predict the magnitude of these anthropogenic factors on coccinellid species worldwide. Moreover, there is a tendency in both research and review to consider each of these drivers independently, even though these drivers do not act in isolation from one another. In our discussion of habitat destruction and fragmentation, we noted that high levels of habitat specialisation and low mobility increase the vulnerability of coccinellids to loss of favourable habitats. In considering climate change, we have noted the importance of north–south wildlife corridors to allow specialised species to maintain environmental constancy. If one considers these two drivers together, it is obvious that habitat specialists, that also have limited climatic niches, are in a very precarious position. The habitats that they currently occupy may well become unfavourable, not due to habitat loss, but through climatic change. However, due to habitat fragmentation, the nearest favourable patch that provides a suitable climatic niche may now be far outside the dispersal range of these species. Add in additional pressures imposed by highly competitive invasive eurytopic species, such as *H. axyridis* and *C. septempunctata*, and pollution effects, and the future for many stenotypic species may be bleak. Strangely, the conclusions from the evidence available in the literature are conflicting. However, this is probably because the vast majority of studies on coccinellids have been on species that are of agricultural importance (providing an ecosystem service of pest control). Many of these species have been chosen specifically because they exhibit behaviours that contribute to their resilience, stability and success as biological control agents (polyphagous, eurytopic predators which exhibit phenotypic plasticity). Most known species of coccinellid have received little or no attention, and at least half of these are habitat specialists with both narrow food ranges, and narrow climatic niches. It is difficult to assess the impact that anthropogenic factors will have on most species of coccinellid but logic suggests that the direct anthropogenic drivers of environmental change, both individually and in concert, will be highly deleterious to all but the most adaptable and eurytopic coccinellids. We conclude with one of the key messages of the Millennium Ecosystem Assessment (2005):

Science can help ensure that decisions are made with the best available information, but ultimately the future of biodiversity will be determined by society.

Acknowledgements HER is supported by the Natural Environment Research Council through the Centre for Ecology & Hydrology. HER is based in the Biological Records Centre (Centre for Ecology & Hydrology) which is co-funded by the Joint Nature Conservation Committee. Professor Michael Majerus, an inspiration to many coccinellid biologists, sadly died in 2009. He will be missed tremendously by many collaborators and friends worldwide. Leaving the final words of this chapter to Mike (Majerus, 1994): *"If my ideas and theories are subsequently put to the test by others who are caught up in the fascination of ladybirds as much as I have been in the last dozen or so years, I will be delighted, whether my ideas are verified or refuted."*

References

Amkar PA (2006) Ecology and biological control application of multicoloured Asian ladybird, *Harmonia axyridis*. A review. Biocontrol Sci Technol 16:111–128

Banks JE (1999) Differential responses of two agroecosystem predators, *Pterostichus madidus* (Coleoptera: Carabidae) and *Coccinella septempunctata* (Coleoptera: Coccinellidae) to habitat composition and fragmentation-scale manipulations. Can Entomol 131:645–657

Barron A, Wilson. K (1998) Overwintering survival in the seven spot ladybird, *Coccinella septempunctata* (Coleoptera: Coccinellidae). Eur J Entomol 95:639–642

Benton AH, Crump AJ (1981) Observations on the spring and summer behavior of the 12-spotted ladybird beetle, *Coleomegilla maculata* (DeGeer) (Coleoptera: Coccinellidae). J N Y Entomol Soc 89:102–108

Brakefield PM (1984) Ecological studies on the polymorphic ladybird *Adalia bipunctata* in the Netherlands. I. Population biology and geographical variation in melanism. J Anim Ecol 53:761–774

Briggs DJ, Cooke GR, Gilbertson DD (1985) The chronology and environmental framework for early man in the Upper Thames Valley. Br Archaeol Rep 137:1–176

Brown MW, Miller SS (1998) Coccinellidae (Coleoptera) in apple orchards of eastern West Virginia and the impact of invasion by *Harmonia axyridis*. Entomol News 109:143–151

Brown PMJ, Adriaens T, Bathon H, Cuppen J, Goldarazena A, Hagg T, Kenis M, Klausnitzer BE, Kovar I, Loomans AJ, Majerus MEN, Nedved O, Pedersen J, Rabitsch W, Roy HE, Ternois V, Zakharov I, Roy DB (2008) *Harmonia axyridis* in Europe: spread and distribution of a non-native coccinellid. BioControl 53:5–22

Chapin B, Brou VA (1991) *Harmonia axyridis* (Pallas), the third species of the genus to be found in the United States (Coleoptera: Coccinellidae). Proc Entomol Soc Wash 93:630–635

Chen FJ, Wu G, Parajulee MN, Ge F (2007) Impact of elevated CO_2 on the third trophic level: a predator *Harmonia axyridis* and a parasitoid *Aphidius picipes*. Biocontrol Sci Technol 17:313–324

Colunga-Garcia M, Gage SH (1998) Arrival, establishment, and habitat use of the multicolored Asian lady beetle (Coleoptera: Coccinellidae) in a Michigan landscape. Environ Entomol 27:1574–1580

Cottrell TE (2007) Predation by adult and larval Lady Beetles (Coleoptera: Coccinellidae) on initial contact with Lady Beetles eggs. Environ Entomol 36:390–401

Cottrell TE, Yeargan KV (1998) Intraguild predation between an introduced lady beetle, *Harmonia axyridis* (Coleoptera: Coccinellidae) and a native lady beetle, *Coleomegilla maculata* (Coleoptera: Coccinellidae). J Kans Entomol Soc 71:159–163

Coviella CE, Trumble JT (1999) Effects of elevated atmospheric carbon dioxide on insect-plant interactions. Conserv Biol 13:700–712

Cronin *JT* (2004) Host-parasitoid extinction and colonization in a fragmented prairie landscape. Oecologia 139:503–514

Davies KF, Melbourne BA, Margules CR (2001) Effects of within and between patch processes on community dynamics in a fragmentation experiment. Ecology 82:1830–1846

De Bach P (1964) Biological control of insect pests and weeds. Chapman and Hall, London

de Jong PW, Brakefield PM (1998) Climate and change in clines for melanism in the two-spot ladybird, Adalia bipunctata (Coleoptera: Coccinellidae). Proc Royal Soc Lond B 265:39–43

Dixon AFG (2000) Insect predator-prey dynamics: ladybird beetles and biological control. Cambridge University Press, Cambridge

Dobrzhanskii FG, Dobzhansky T (1922) Imaginal diapause in Coccinellidae. Izvestiia Otdel Prikladnaia Entomologia 2:103–124

Dobson A, Lodge D, Alder J, Cumming GS, Keymer J, McGlade J, Mooney H, Rusak JA, Sala O, Wolters V, Wall D, Winfree R, Xenopoulos MA (2006) Habitat loss, trophic collapse and the decline of ecosystem services. Ecology 87:1915–1924

Elliott N, Kieckhefer R, Kauffman W (1996) Effects of an invading coccinellid on native cocccinellids in an agricultural landscape. Oecologia 105:537–544

Evans EW (2000) Morphology of invasion: body size patterns associated with establishment of *Coccinella septempunctata* (Coleoptera: Coccinellidae) in western North America. Eur J Entomol 97:469–474

Ewert MA, Chiang HC (1966) Effects of some environmental factors on the distribution of three species of Coccinellidae in their microhabitat. In: Hodek I (ed) Ecology of aphidophagous insects. Academia, Prague and Dr W. Junk, The Hague, pp 195–219

Felix S, Soares AO (2004) Intraguild predation between the aphidophagous ladybird beetles *Harmonia axyridis* and *Coccinella undecimpunctata* (Coleoptera: Coccinellidae): the role of body weight. Eur J Entomol 101:237–242

Ferran A, Gubanti P, Iperti G, Migeon A, Onillon J (1989) La repartition spatiale des différents stades de *Coccinella septempunctata* dans un champ de blé: variation au cours de la saison. Entomol Exp Appl 53:229–236

Galvan TL, Koch RL, Hutchison WD (2005) Toxicity of commonly used insecticides in sweet corn and soybean to Multicoloured Asian Lady Beetle (Coleoptera: Coccinellidae). J Econ Entomol 98:780–789

Gendebien A, Carlton-Smith C, Izzo M, Hall J (1999) UK sewage sludge survey. Environment Agency Research and Development Technical Report P165, Bristol

Gordon RD (1985) The Coccinellidae (Coleoptera) of America North of Mexico. J N Y Entomol Soc 93:1–912

Grafton-Cardwell EE, Gu P (2003) Conserving Vedalia Beetle, *Rodolia cardinalis* (Mulsant) (Coleoptera: Coccinellidae), in citrus: a continuing challenge as new insecticides gain registration. J Econ Entomol 96:1388–1398

Green ID, Merrington G, Tibbett M (2003) Transfer of cadmium and zinc from sewage sludge amended soil through a plant-aphid system to newly emerged adult ladybirds (*Coccinella septempunctata*). Agric Ecosyst Environ 99:171–178

Hagen KS (1962) Biology and ecology of predaceous Coccinellidae. Ann Rev Entomol 7:289–326

Hironori Y, Katsuhiro S (1997) Cannibalism and interspecific predation in two predatory ladybirds in relation to prey abundance in the field. Entomophaga 42:153–163

Hodek I (1996a) Food relationships. In: Hodek I, Honěk A (eds) Ecology of coccinellidae. Kluwer Academic Publishers, Dortrecht, pp 143–238

Hodek I (1996b) Dormancy. In: Hodek I, Honěk A (eds) Ecology of coccinellidae. Kluwer Academic Publishers, Dortrecht, pp 239–318

Hodek I, Cerkasov J (1963) Imaginal dormancy in *Semiadalia undecimnotata* Schneid. (Coccinellidae, Col.) II. Changes in water, fat and glycogen content. Vestnik Ceskoslovenske Zoologicke Spolecnosti 27:298–318

Honěk A (1979) Plant density and occurrence of *Coccinella septempunctata* and *Propylea quatuordecimpunctata* (Coleoptera, Coccinellidae) in cereals. Acta Entomol Bohemoslov 76:308–312

Honěk A (1985) Habitat preferences of aphidophagous coccinellids [Coleoptera]. Entomophaga 30:253–264

Honěk A, Hodek I (1996) Distribution in habitats. In: Hodek I, Honěk A (eds) Ecology of coccinellidae. Kluwer Academic Publishers, Dortrecht, pp 95–142

Howland DE, Hewitt GM (1995) Phylogeny of the Coleoptera based on mitochondrial cytochrome oxidase 1 sequence data. Insect Mol Biol 4:203–215

Hurst GDD, Sharpe RG, Broomfield AH, Walker LE, Majerus TMO, Zakharov IA, Majerus MEN (1995) Sexually transmitted disease in a promiscuous insect, *Adalia bipunctata*. Ecol Entomol 20:230–236

Kajita Y, Takano F, Yasuda H, Agarwala BK (2000) Effects of indigenous ladybird species (Coleoptera: Coccinellidae) on the survival of an species in relation to prey abundance. Appl Entomol Zool 35:473–479

Kajita Y, Yasuda H, Evans EW (2006) Effects of native ladybirds on oviposition of the exotic species, *Adalia bipunctata* (Coleoptera: Coccinellidae), in Japan. Appl Entomol Zool 41:57–61

Kareiva P (1987) Habitat fragmentation and the stability of predator-prey interactions. Nature 326:388–390

Koch RL (2003) The multicoloured Asian lady beetle, *Harmonia axyridis*: a review of its biology, uses in biological control and non-target impacts. J Insect Sci 3:1–16

Koch RL, Burkness EC, Wold SJ, Hutchison WD (2004) Phytophagous preferences of the multicolored Asian lady beetle (Coleoptera: Coccinellidae) to autumn ripening fruit. J Ecol Entomol 97:539–544

Koch RL, Galvan TL (2008) Bad side of a good beetle: the North American experience with *Harmonia axyridis*. BioControl 53:23–36

Kovach J (2004) Impact of the multicolored Asian lady beetle as a pest of fruit and people. Am Entomol 50:165–167

LaMana ML, Miller JC (1996) Field observation on *Harmonia axyridis* (Pallas) (Coleoptera: Coccinellidae) in Oregon. Biol Control 6:232–237

Laurance WF, Cochrane MA, Bergen S, Fearnside PM, Delamonica P, Barber C, D'Angelo S, Fernandes T (2001) The future of the Brazilian Amazon. Science 291:438–439

Lövei GL, Sarospataki M, Radwan ZA (1991) Structure of ladybird (Coleoptera: Coccinellidae) assemblages in apple: changes through developmental stages. Environ Entomol 20:1301–1308

Majerus MEN (2003) Ladybugs. In: Resh VH, Cardé RT (eds) Encyclopedia of insects. Academic Press, San Diego, pp 618–622

Majerus MEN (1994) Ladybirds. No. 81, new naturalist series. HarperCollins, London

Majerus MEN, Roy HE, Brown PMJ, Ware RL, Shields C (2006a) Guide to ladybirds of the British isles. Field Studies Council Publications, Shrewsbury

Majerus MEN, Strawson V, Roy HE (2006b) The potential impacts of the arrival of the Harlequin ladybird, *Harmonia axyridis* (Pallas) (Coleoptera: Coccinellidae), in Britain. Ecol Entomol 31:207–215

McGarigal K, Cushman SA (2002) Habitat fragmentation: a reassessment. Ecol Appl 12:335–345

Michaud JP (2001) Responses of two lady beetles to eight fungicides used in Florida citrus: implications for biological control. J Insect Sci 1:1–6

Michaud JP (2002a) Invasion of the Florida citrus ecosystem by *Harmonia axyridis* (Coleoptera: Coccinellidae) and asymmetric competition with a native species, *Cycloneda sanguinea*. Environ Entomol 31:827–835

Michaud JP (2002b) Non-target impacts of acaricides on ladybeetles in citrus: a laboratory study. Fla Entomol 85:191–196

Michaud JP (2002c) Relative toxicity of six insecticides to *Cycloneda sanguniea* and *Harmonia axyridis* (Coleoptera: Coccinellidae). J Entomol Sci 37:83–93

Michaud JP, Grant AK (2003) Sub-lethal effects of a copper sulfate fungicide on development and reproduction in three coccinellid species. J Insect Sci 3:16. Available online: insectscience.org/3.16

Millenium Ecosystem Assessment (2005) Ecosystems and human well-being: biodiversity synthesis. World Resources Institute, Washington, DC

Müller CB, Godfray HCJ (1997) Apparent competition between two aphid species. J Anim Ecol 66:57–64

Nault BA, Kennedy GG (2003) Establishment of multicoloured Asian lady beetle in Eastern North Carolina: seasonal abundance and crop exploitation within an agricultural landscape. BioControl 48:363–378

Nriagu JO, Pacyna JM (1988) Quantitative assessment of worldwide contamination of air, water and soils by trace metals. Nature 333:134–139

Ochou OG, Martin T (2003) Activity spectrum of spinosad and indoxacarb: rationale for an innovative pyrethroid resistance management strategy in West Africa. Resistant Pest Manage Newsl 12:75–81

Osborne PJ (1971) The insect faunas of the Wandle Gravels. Proc Trans Croydon Nat Hist Soc 14:162–175

Parmesan CN, Yohe G (2003) A globally coherent fingerprint of climate change impacts across natural systems. Nature 421:37–42

Pell JK, Baverstock J, Roy HE, Ware RL, Majerus MEN (2008) Intraguild predation involving *Harmonia axyridis*: a review of current knowledge and future perspectives. BioControl 53: 147–168

Phoofolo MW, Obrycki JJ (1998) Potential for intraguild predation and competition among predatory Coccinellidae and Chrysopidae. Entomol Exp Appl 89:47–55

Qi B, Gordon G, Gimme W (2001) Effects of neem-fed prey on the predacious insects *Harmonia conformis* (Boisduval) (Coleoptera: Coccinellidae) and *Mallada signatus* (Schneider) (Neuroptera: Chrysopidae). Biol Control 22:185–190

Rand TA, Tscharntke T (2007) Contrasting effects of natural habitat loss on generalist and specialist aphid natural enemies. Oikos 116:1353–1362

Rand TA, Tylianakis JM, Tscharntke T (2006) Spillover edge effects: the dispersal of agriculturally subsidised insect natural enemies into adjacent natural habitats. Ecol Lett 9:603–614

Roy HE, Brown PMJ, Majerus MEN (2006) *Harmonia axyridis*: a successful biocontrol agent or an invasive threat? In: Eilenberg J, Hokkanen H (eds) An ecological and societal approach to biological control. Kluwer Academic Publishers, Netherlands

Sakuratani Y, Marsumoto Y, Oka M, Kubo T, Fuji A, Uotani M, Teraguchi. T (2000) Life history of *Adalia bipunctata* (Coleoptera: Coccinellidae) in Japan. Eur J Entomol 97:555–558

Salt G (1920) A contribution to the ethology of the Meliponinae. Trans Entomol Soc Lond 77: 431–470

Sato S, Dixon AFG, Hironori Y (2003) Effect of emigration on cannibalism and intraguild predation in aphidophagous ladybirds. Ecol Entomol 28:628–633

Sloggett JJ (2005) Are we studying too few taxa? Insights from aphidophagous ladybird beetles (Coleoptera: Coccinellidae). Eur J Entomol 102:391–398

Sloggett JJ, Majerus MEN (2003) Adaptations of *Coccinella magnifica*, a myrmecophilous coccinellid to aggression by wood ants (*Formica rufa* group). II. Larval behaviour, and ladybird oviposition location. Eur J Entomol 100:337–344

Smith BC (1971) Effects of various factors on the local distribution and density of coccinellid adults on corn (Coleoptera, Coccinellidae). Can Entomol 103:1115–1120

Snyder WE, Clevenger GM, Eigenbrode SD (2004) Intraguild predation and successful invasion by introduced ladybird beetles. Oecologia 140:559–565

Snyder WE, Evans EW (2006) Ecological effects of invasive arthropod generalist predators. Ann Rev Ecol Evol Syst 37:95–122

Southwood TRE (1977) Habitat the templet for ecological studies? J Anim Ecol 46:337–365

Stacey DA, Fellowes MDE (2002) Influence of elevated CO_2 on interspecific interactions at higher trophic levels. Glob Chang Biol 8:668–678

Stoner KJL, Joern A (2004) Landscape versus local habitat scale influences to insect communities from tallgrass prairie remnants. Ecol Appl 14:1306–1320

Völkl W (1995) Behavioural and morphological adaptations of the coccinellid, *Platynaspis luteorubra* for exploiting ant-attending resources (Coleoptera: Coccinellidae). J Insect Behav 8:273–281

Ware RL, Majerus MEN (2008) Intraguild predation of immature stages of British and Japanese coccinellids by the invasive ladybird *Harmonia axyridis*. BioControl 53:169–188

Webberley KM, Tinsley M, Sloggett JJ, Majerus MEN, Hurst GDD (2006) Spatial variation in the incidence of sexually transmitted parasites of the ladybird beetle *Adalia bipunctata*. Eur J Entomol 103:793–797

Welch VL, Sloggett JJ, Webberley KM, Hurst GDD (2001) Short-range clinal variation in the prevalence of a sexually transmitted fungus associated with urbanisation. Ecol Entomol 26:547–550

Woodroffe R, Ginsberg JR (1998) Edge effects and the extinction of populations inside protected areas. Science 280:2126–2128

Yasuda H, Evans EW, Kajita Y, Urakawa K, Takizawa T (2004) Asymmetrical larval interactions between introduced and indigenous ladybirds in North America. Oecologia 141:722–731

Yasuda H, Shinya Y (1997) Cannibalism and interspecific predation in two predatory ladybirds in relation to prey abundance in the field. Entomophaga 42:153–163

Youn YN, Seo MJ, Shin JG, Jang C, Yu YM (2003) Toxicity of greenhouse pesticides to multicolored Asian lady beetles, *Harmonia axyridis* (Coleoptera: Coccinellidae). Biol Control 28:164–170

Chapter 10
Ecological Costs of Alarm Signalling in Aphids

Yannick Outreman, Grit Kunert, Jean-Christophe Simon, and Wolfgang W. Weisser

Abstract The existence of costs is implicit in the evolution of inducible defences against natural enemies yet attempts to demonstrate such costs have often proved elusive. Ecological costs occur when defence expression results in a change in biotic interactions such that the fitness of the individual expressing the trait is reduced. Aphids have evolved a particular form of inducible anti-predator behaviour that involves the emission of alarm pheromone. This pheromone, whose main component is (E)-β-Farnesene (EBF), is released by individuals attacked by a natural enemy and initiates behavioural responses in other aphid colony members such as escape reactions. We placed colonies of about 250 pea aphids in the field and exposed them daily to artificial EBF or control solvent (hexane). Aphid survival in the experiment was very low mainly due to predation by ants and colonies exposed to artificial EBF showed a reduced survival time (2.35±0.23 days; mean±s.e.) compared to the control treatment (3.51±0.42 days). Our results show that alarm signalling in aphids is associated with the ecological cost of attracting additional natural enemies and demonstrate that a full understanding of the evolution of inducible defences has to consider a species' complete network of ecological interactions. As EBF, like many other volatile organic compounds mediating organismic interactions, breaks down more quickly at higher ozone levels, global climatic change may result in a change in these interactions.

Keywords (*E*)-β-farnesene · Ant predators · Multitrophic interactions · Pea aphid · Alarm pheromone

---

Y. Outreman (✉)
UMR 1099 INRA-Agrocampus Ouest-Université Rennes 1 "Biologie des Organismes et des Populations appliquée à la Protection des Plantes" [BIO3P], Le Rheu Cedex, France; Institute of Ecology, Friedrich-Schiller-University, 07743 Jena, Germany
e-mail: yannick.outreman@rennes.inra.fr

P. Kindlmann et al. (eds.), *Aphid Biodiversity under Environmental Change*,
DOI 10.1007/978-90-481-8601-3_10, © Springer Science+Business Media B.V. 2010

10.1 Introduction

Inducible defences have been described for many plant-herbivore and predator-prey systems, and studies have mostly demonstrated that these adaptations can reduce the risk of being eaten by a natural enemy (Karban and Baldwin 1997; Tollrian and Harvell 1998). The evolution of an inducible defence also implies, however, that there are associated costs which could be saved at times of little or no predation risk (Tollrian and Harvell 1998). Nevertheless, for many systems it has proven quite difficult to demonstrate the existence of such costs (e.g. Agrawal 1998; Agrawal and Rutter 1998; Baldwin 1998; Cipollini 2002; Heil and Baldwin 2002; Tian et al. 2003). One reason for this difficulty is that costs can take many forms. Costs can arise directly from the expression of the defence, or indirectly via interactions with other species. Direct costs include physiological allocations costs (Karban and Baldwin 1997) and auto-toxicity (Strauss et al. 2002). Indirect, or ecological costs occur when defence expression results in reduced pollination, attracts enemies, or impairs expression of other resistance traits (Heil and Baldwin 2002; Cipollini et al. 2003). The empirical measurement of ecological costs requires an intimate knowledge of the interactions of the target species with its competitors, natural enemies and mutualistic partners in the field, a comprehensive task if the species are embedded in a complex interaction web (e.g., Adler et al. 2001). As a consequence of the complexity of interaction webs, it is likely that many ecological costs of inducible defences remain to be discovered.

Aphids (Sternorryncha: Aphidoidea) are attacked by a wide range of natural enemies such as parasitoid wasps, ladybirds, lacewings and hoverfly larvae which strongly influence the growth and persistence of aphid colonies (Bänsch 1964; Minks and Harrewijn 1988; Dixon 1998). Under enemy attack, aphids secrete droplets from their siphunculi, a pair of tube-like structures on the posterior dorsal surface of the abdomen. In addition to the direct defensive function of gluing or "waxing" predator mouthparts or parasitoid antennae, these droplets contain an alarm pheromone whose main and sometimes only component is the sesquiterpene (*E*)-β-farnesene, EBF (Bowers et al. 1972; Kislow and Edwards 1972; Edwards et al. 1973; Nault et al. 1973; Pickett and Griffiths 1980). EBF triggers various behavioural reactions in aphids of the attacked colony such as increased alertness, withdrawal of stylets from the plant, walking and dropping off the host plant, any of which may reduce the risk of being attacked by the enemy (Montgomery and Nault 1977; Wohlers 1980; Dixon 1998). In pea aphids, EBF is also involved in triggering the production of winged morphs in the presence of predators (Kunert et al. 2005). Thus, EBF functions in defensive communication among aphids and induces escape responses that reduce predation risk. The costs of EBF release appear relatively clear-cut: aphids incur opportunity costs leaving good feeding sites, increase their risk of starvation and possibly their risk of predation by ground-foraging predators (Roitberg and Myers 1979; Dill et al. 1990; Stadler et al. 1994; Losey and Denno 1998). In addition, there are physiological costs of EBF secretion: secreting aphids delay the production of offspring compared to non-secreting individuals (Mondor and Roitberg 2003). Thus, any non-essential or inappropriate release of EBF would clearly be maladaptive. However, because EBF-release is very efficient in reducing the immediate risk of predation for those individuals that quickly respond by dropping from the plant, it is generally assumed that the benefits of its emission largely outweigh the costs.

When aphids release EBF in the presence of a natural enemy, the information conveyed is not only that aphids are being attacked, but also more generally that there are aphids available on a plant. For this reason a number of studies have investigated the attractiveness of EBF to aphid natural enemies (Hatano et al. 2008). Laboratory experiments with ladybirds (Zhu et al. 1999; Al Abassi et al. 2000), lacewings (Boo et al. 1998; Zhu et al. 1999) and aphid parasitoids (Grasswitz and Paine 1992; Micha and Wyss 1996) have provided some evidence that aphid natural enemies are attracted by, or at least show antennal responses to EBF. Nevertheless, Nakamuta (1991) did not find that presence of EBF influenced ladybird foraging paths and more recently, Mondor et al. (2000) found no evidence for attraction of the ladybird *Harmonia axyridis* to aphid alarm pheromone. Thus, at present there is only weak laboratory evidence for a cost of EBF release to aphid colonies due to attraction of further natural enemies, and such ecological costs have not yet been observed in the field (Hatano et al. 2008).

For this chapter, the effect of aphid alarm pheromone emission on the fate of colonies of pea aphid, *Acyrthosiphon pisum*, was tested in the field. EBF is the only component of pea aphid alarm pheromone and individuals show strong behavioural reactions in the presence of synthetic EBF (Kunert et al. 2005). We ask the following questions: (1) Is there a measurable effect of EBF emission on aphid colony performance under field conditions? (2) Is there evidence for EBF-mediated attraction of natural enemies to aphid colonies? (3) Are there costs of aphid alarm signalling that need to be accounted for in our understanding of the evolution of this inducible response? Finally, we will discuss our findings in the light of global climatic change.

10.2 Materials and Methods

10.2.1 Aphids and Colony Production

We used a red clone of the pea aphid that was originally collected in Bayreuth, Germany and that has been used previously for alarm pheromone experiments (Kunert et al. 2005). Aphids were reared on broad beans, *Vicia faba*, L. (variety "The Sutton", Nickerson-Zwaan, UK). The plants were grown in 10 cm-diameters pots, covered with air-permeable cellophane bags to prevent aphid escape. For the experiment, aphid colonies were established by placing three apterous adults on a broad bean plant and allowing them to reproduce over 4 days before removing them from the plant. When their progeny (approximately 60 individuals per plant) reached the fourth instar, groups of 10 offspring were transferred separately to new broad bean plants to reduce crowding. These nymphs were reared until adulthood and permitted to reproduce for 3 days before use in the experiment. All rearing was done in climate rooms at 20°C, 70±10% relative humidity, and 16L:8D photoperiod.

10.2.2 Experimental Design

The colonies of pea aphid were introduced to the experimental site of the Jena grassland biodiversity experiment (Roscher et al. 2005) on 3.7.2005 at 18:00 h. Colony size was about 250–300 individuals when the experiment started. Plants were placed along the margin of the site next to ruderal vegetation that was up to 1 m high. The distance between paired plants was 1 m and distance between adjacent pairs was about 30 m. A total of 30 pairs of infested plants were set out, one of each pair randomly allocated to the EBF treatment with the other serving as the control.

For the EBF-treatment, 3 μl of a solution of EBF dissolved in *n*–hexane (containing 1,000 ng of EBF) was applied with a micropipette onto a piece of filter paper fixed by a wooden toothpick at the base of the pot, starting on 4.7.2005. Due to short half-life of EBF in air, the solution was applied twice daily, once in the morning starting at 9:00 h and again in the late afternoon. For control plants, 3 μl of the solvent was applied at the same times. The treatments were repeated daily until 10.7.2005. Slug pellets (active ingredient metaldehyde) were applied around all plant plots to prevent herbivory by slugs on the bean plants.

Before each application of the solution, the number of aphids on the plant was counted. In addition, each plant was inspected for the presence of natural enemies, i.e. ants, hoverfly, lacewing or ladybird larvae, anthocorid bugs, parasitoids etc. After the solution was applied, we noted if aphids walked or dropped off the plant. Handling time, i.e. observations and application of EBF or solvent, lasted 2–4 min on each occasion. The experiment continued for 8 days until 11.7.2005, when aphids were counted the last time at 12.00 a.m.

10.2.3 Statistical Analysis

Our data set consisted of a correlated time series (Liang and Zeger 1986) of daily aphid counts as a function of block (plant pair) and treatment (alarm pheromone treatment). Generalised estimating equations (G.E.E.) were used to estimate parameters for the correlated data, assuming a Poisson error distribution with a log link function (the R statistical package "geepack" (Ihaka and Gentleman 1996)).

The effect of the alarm pheromone treatment was analysed using Cox's proportional hazards model (Collett 1994) to estimate the influence of block and treatment on the probability per unit time. The Cox's model is formulated in terms of hazard rate and takes the form

$$h(t,z) = h_0(t)\exp\left\{\sum_{i=1}^{p} \beta_i z_i(t)\right\}$$

where t is the time in days elapsed since the aphid colony was introduced in the field, $h(t,z)$ is the hazard rate (the probability per unit time that an aphid colony becomes extinct), $h_0(t)$ is the baseline hazard, and β_i are the regression coefficients that give the relative contributions of the p covariates $z_i(t)$. The hazard ratio $\exp(\beta_i z_i)$ is a quantitative measure of the effect of each particular covariate. If it is greater than one, the aphid colony extinction risk is increased compared to the baseline (control); if it is smaller than one, the extinction risk is reduced. The effect of each covariable (i.e., treatment and block) and the significance of the overall model were assessed using likelihood ratio tests (Collett 1994). These analyses were carried out using S-plus statistical software (MathSoft, Cambridge, Massachusetts, USA) (Venables and Ripley 1997). All averages are given as mean ± sem.

10.3 Results

Aphid populations declined quickly (GEE: regression coefficient 0.052 ± 0.013, $p < 0.001$, Fig. 10.1) and were greatly reduced even after 1 day on both treatment and control plants. Populations in the EBF treatment declined faster than populations in the control (GEE regression coefficient –0.462 ± 0.223, $P < 0.05$, Fig. 10.1). As a consequence of the fast decline in aphid numbers, the survival of aphid colonies was short. Six pairs of plants had to be replaced on the first application date because of extinction of at least one colony of the pair prior to treatment. At the end of the experiment, only a single aphid was still alive. The fitted proportional hazard model showed that the EBF treatment increased the extinction risk of an aphid colony by the factor 1.624 (regression coefficient –0.485 ± 0.195, $\chi^2 = 5.96$, df. = 1, $P < 0.05$, Fig. 10.2). Hence, colonies receiving the EBF treatment survived for a shorter time (2.35 ± 0.23 days) than control ones (3.51 ± 0.42 days, Fig. 10.2). The extinction risk also varied among the different plant pairs ($\chi^2 = 49.21$, df =27, $P < 0.01$).

Aphids were never observed to drop or walk off the plant following application of EBF. Natural enemies were observed on at least one plant in 17 of the 30 plant

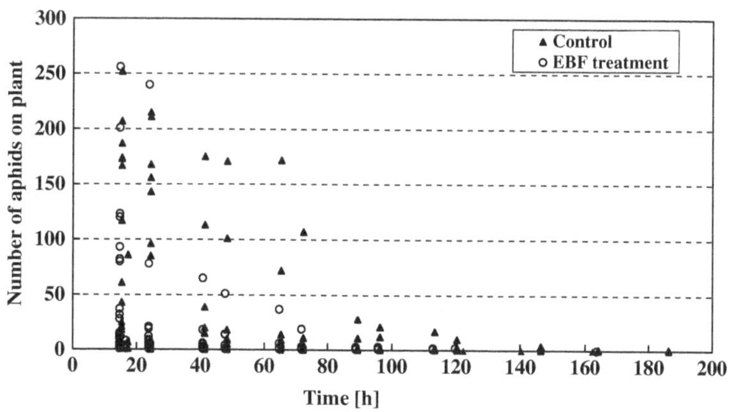

Fig. 10.1 Aphid colony size as a function of treatment over the time course of the experiment. Initial colony size on day 0 (t=0 h) was 250–300 aphids on each plants. N=60 plants. The time of inspection of colonies has been slightly changed to improve visibility (alarm pheromone (EBF) treatment –0.3 h, control +0.3 h)

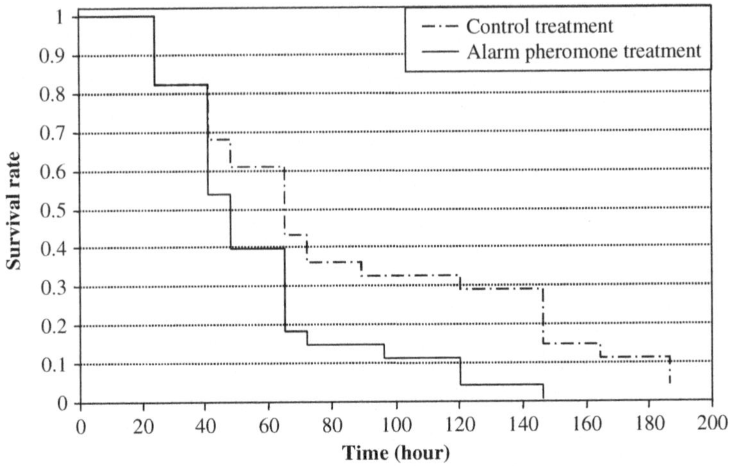

Fig. 10.2 Estimated survival curves of pea aphid colonies exposed to alarm pheromone or solvent (Kaplan-Meier estimates)

pairs, and at least once during the course of the experiment on 14 of the 30 plants of the EBF treatment and 10 of the 30 control plants. Even on the first day, a number of aphid predators were observed in aphid colonies, most notably a large number of ants (*Lasius niger*) that were carrying aphids off the plants. Ants were observed during 19 inspections remaining aphid colonies whereas other natural enemies were less common (predatory Heteroptera and spiders were observed during four inspections each, and parasitoid wasps on two inspections (Table 10.1)). There were no significant differences in the occurrence of natural enemies on EBF treatment and control plants, likely because of the low numbers of observations (sign test, $P > 0.5$ for each natural enemy).

Table 10.1 Occurrence of natural enemies in aphid colonies. Shown are the numbers of pairs where a particular predator occurred first on the plant of a particular treatment. Because in some pairs several natural enemies were observed, the total number of pairs listed here (23) is higher than the number of pairs where predators were observed (17)

Natural enemy	Ants	Spiders	Parasitoids	Heteroptera	All
First observed in EBF-treatment	6	3	2	2	13
First observed in control	3	1	1	2	7
First occurrence simultaneous in both treatments	3	0	0	0	3

10.4 Discussion

In our experiment, survival of most aphid colonies was very low. While the strong decline in aphid numbers (from 250 to 300 individuals to zero individuals in the time-course of a few days) was surprising to us, such low survival of aphids has in fact been reported in other aphid species where individual colonies were tagged and followed over time (e.g. Addicott 1978; Weisser 2000). The decrease in aphid colony sizes did not appear to be due to a sudden decrease in plant suitability, as no wilting, drying or plant damage was observed during the experiment, nor due to adverse climatic conditions (i.e., no rainfall during the experiment). Neither was the decline due to emigration of aphids in response to perception of EBF, as neither dropping nor walking was observed after application. One striking observation was the large number of ants that carried pea aphids off plants, suggesting that predation by ants was responsible for the fast decline in aphid numbers, and for the differences between the EBF and control colonies. Our results suggest that, aside from the benefit of cuing escape from predators, the emission of alarm pheromone may also carry a large ecological cost for aphids through the attraction of natural enemies.

A relationship between aphid alarm pheromone and ant aggression has been reported before, albeit with a focus on aphid-ant mutualism. In a comprehensive study, Nault et al. (1976) used a laboratory setting to expose several myrmecophilous and non-myrmecophilous aphid species to both an ant colony and EBF. In myrmecophilic aphids, ants became very aggressive in the presence of alarm pheromone and increased their rate of attack on aphid predators, but did not attack aphids. However, when alarm pheromone was applied to colonies of aphid species that are not ant-tended, ants became aggressive towards the aphids and sometimes carried them off the plant (Nault et al. 1976). The studies by Nault et al. (1976) emphasize that (1) EBF is perceived by ants and elicits a response and (2) that EBF can act as a stimulus that increases aggression of ants towards non-myrmecophilic aphids such as the pea aphid. In our experiment, we observed heavy predation on pea aphid colonies by groups of ant workers, especially in the beginning of the experiment. Although we also observed other predators such as ladybirds and anthocorid bugs that may also have caused aphid mortality, ants were the only predator that was observed in large numbers on the plants. Because of their recruitment behaviour, ants are able to reduce aphid colony size from almost 300 to near zero within a few hours, as was observed in our experiment.

Ants did not only forage in colonies of the EBF treatment, but the regular patrolling of workers probably resulted in ant predation in almost all colonies, although this could not always be documented due to the short inspection periods. In fact, at first sight, the hypothesis that ant attraction and/or aggression was higher in the EBF-treatment compared to the control appears to be unsupported by our data on first occurrence of ants in aphid colonies (Table 10.1); while ants were indeed more often observed first in EBF colonies, the time of first observation of ants in a colony did not differ between treatments. These observations suffer, however, from the problem that in many pairs no natural enemies were observed at all, due to the short observation intervals (we infer that many short predator visits occurred when the observer was not present), and to the fact that predators were rarely observed on aphid-free plants. Thus, in plant pairs where aphids went extinct first in the EBF treatment and where predation happened between two observation intervals, it was possible that ants were recorded first in the control treatment. This is because the aphid-free plant in the EBF treatment would not continue to attract ants while in the control treatment ants would continue to prey on the remaining aphids (Table 10.1). Thus, although it was not possible to directly link each extinction to the presence of particular predators, the available evidence strongly suggests that predation, in particular by ants, led to the fast demise of aphid colonies which happened faster in the EBF treatment.

These results have implications for our understanding of the cost and benefits of defence communication in insects. Our results suggest that release of the aphid alarm pheromone can result in significant costs to the aphids in situations where predators are in the vicinity that may detect the alarm communication. Because ant workers that encounter a food resource recruit other workers, it is conceivable that even a single dose of EBF may be enough to attract a large number of ants that can carry off aphids in a colony within a single day. As a consequence, there should be a selective advantage for aphids to minimize EBF emission to situations where the benefits of alarm communication, i.e. reduction in predation risk from a predator already on the plant, are higher than the costs of releasing EBF, i.e. attraction of further natural enemies. However, nothing is known about the ability of aphids to adjust the amount of EBF given off in response to attack by predators. There is evidence that the amount of EBF emitted by aphids is much lower than the amount of EBF present in the reservoirs next to the base of the siphunculi (Mondor et al. 2000; Kunert et al. 2005), but no explanation has been suggested why aphids do not empty the entire reservoir when attacked, given that attacked individuals usually die from the attack. From a theoretical point of view, an adjustment of EBF emission to the minimum level needed to warn conspecifics (clones) would be beneficial if this reduces the probability of attracting natural enemies. Incidentally, the importance of ants for aphid colony survival in our experiment also gives support to the argument that in the evolution of ant-attendance those aphid species that form mutualistic relationships with ants may not just have recruited ants to reduce the risk of predation by other natural enemies, but to reduce predation by the ants themselves (Stadler et al. 2001; 2002; 2003).

EBF like many other volatile organic compounds reacts with ozone and is degraded in the process (Pinto et al. 2007; Yuan et al. 2009). For this reason, and because of the small amounts emitted, it is likely that EBF is not important for long-range attraction of predators but it only detected when a predator is near a plant or already on the plant (Hatano et al. 2008). This is a likely reason why in the current experiment ants played such an important role. As ant workers regularly patrol plants there is a certain probability that an ant is present on a plant when an aphid emits EBF, for example due to attack by another predator. An EBF emission event is therefore more likely to alert ant workers than to attract more individuals of the predator species that attacked the aphid. With an increase in ozone concentrations under global climatic change it is likely that the lifetime of EBF, and hence the distance it is carried through the air, will decrease in areas of high ozone, and during daytimes when ozone concentrations peak. How will this affect aphid alarm communication and the interaction between predator and prey? As climatic changes implies a number of changes such as an increase in temperature, a decrease in precipitation or an increase in CO_2, all of which may also affect emission, lifetime and detectibility of volatile organic compounds, the net result is difficult to predict (Yuan et al. 2009). Nevertheless, it appears that the increased aerial deterioration of EBF will further reduce the likelihood that predators are attracted to plants with aphids. For aphid-aphid communication, and for the interaction between aphid and ants, there may be little change unless reduction in EBF lifetime is such that even the distance from one leaf to another exceeds the travel distance of the compound. This seems to be an unlikely scenario but measurements of EBF travel distances under various ozone concentrations are needed to quantitatively understand the consequences of global climatic change for communication in the aphid ecological network.

Acknowledgements We thank Ingrid Jakobi for help with rearing plants and animals. This study was supported by the German Academic Exchange Service (DAAD) within the framework of a PPP-Procope German-French collaboration program.

References

Addicott JF (1978) The population dynamics of aphids on fireweed: a comparison of local populations and metapopulations. Can J Zool 56:2554–2564

Adler LS, Karban R, Strauss SY (2001) Direct and indirect effects of alkaloids on plant fitness via herbivory and pollination. Ecology 82:2032–2044

Agrawal AA (1998) Induced responses to herbivory and increased plant performance. Science 279:1201–1202

Agrawal AA, Rutter MT (1998) Dynamic anti-herbivore defense in ant-plants: the role of induced responses. Oikos 83:227–236

Al Abassi S, Pettersson J, Pickett JA, Wadhams LJ, Woodcock CM (2000) Response of the seven-spot ladybird to an aphid alarm pheromone and an alarm pheromone inhibitor is mediated by paired olfactory cells. J Chem Ecol 26:1765–1771

Baldwin IT (1998) Jasmonate-induced responses are costly but benefit plants under attack in native populations. Proc Nat Acad Sci USA 95:8113–8118

Bänsch R (1964) Vergleichende Untersuchungen zur Biologie und zum Beutefangverhalten aphidivorer Coccinelliden, Chrysopiden und Syrphiden. Zoologisches Jahrbuch der Systematik 91:271–340

Boo KS, Chung IB, Han KS, Pickett JA, Wadhams LJ (1998) Response of the lacewing *Chrysopa cognata* to pheromones of its aphid prey. J Chem Ecol 24:631–643

Bowers WS, Nault LR, Webb RE, Dutky SR (1972) Aphid alarm pheromone: isolation, identification, synthesis. Science 177:1121–1122

Cipollini DF (2002) Does competition magnify the fitness costs of induced responses in *Arabidopsis thaliana*? A manipulative approach. Oecologia 131:514–520

Cipollini D, Purrington CB, Bergelson J (2003) Costs of induced responses in plants. Basic Appl Ecol 4:79–85

Collett D (1994) Modeling survival data in medical research. Chapman & Hall, London

Dill LM, Frazer BD, Roitberg BD (1990) The economics of escape behaviour in the pea aphid, *Acyrthosiphon pisum*. Oecologia 83:473–478

Dixon AFG (1998) Aphid ecology. Chapman & Hall, London
Edwards LJ, Sidall JB, Dunham LL, Dunhall LL, Uden P, Kislow CJ (1973) Trans-Beta-Farnesene, alarm pheromone of the green peach aphid, *Myzus persicae* (Sulzer). Nature 241:126–127
Grasswitz TR, Paine TD (1992) Kairomonal effect of an aphid cornicle secretion on *Lysiphlebus testaceipes* (Cresson) (Hymenoptera, Aphidiidae). J Insect Behav 5:447–457
Hatano E, Kunert G, Michaud J, Weisser W (2008) Chemical cues mediating aphid location by natural enemies. Eur J Entomol 105:797–806
Heil M, Baldwin IT (2002) Fitness costs of induced resistance: emerging experimental support for a slippery concept. Trends Plant Sci 7:61–67
Ihaka R, Gentleman R (1996) R: a language for data analysis and graphics. J Comput Graph Stat 5:299–314
Karban R, Baldwin IT (1997) Induced responses to herbivory. The University of Chicago Press, Chicago
Kislow C, Edwards LJ (1972) Repellent odour in aphids. Nature 235:108–109
Kunert G, Otto S, Röse USR, Gershenzon J, Weisser WW (2005) Alarm pheromone mediates production of winged dispersal morphs in aphids. Ecol Lett 8:596–603
Liang KY, Zeger SL (1986) Longitudinal data analysis using generalized linear models. Biometrika 73:13–22
Losey JE, Denno RF (1998) Interspectific variation in the escape responses of aphids: effect on risk of predation from foliar-foraging and ground-foraging predators. Oecologia 115:245–252
Micha SG, Wyss U (1996) Aphid alarm pheromone (E)-beta-farnesene: a host finding kairomone for the aphid primary parasitoid Aphidius uzbekistanicus (Hymenoptera: Aphidiinae). Chemoecology 7:132–139
Minks AK, Harrewijn P (1988) Aphids, their biology, natural enemies and control, vol. B. Elsevier, Amsterdam
Mondor EB, Baird DS, Slessor KN, Roitberg BD (2000) Ontogeny of alarm pheromone secretion in pea aphid, *Acyrthosiphon pisum*. J Chem Ecol 26:2875–2882
Mondor EB, Roitberg BD (2003) Age-dependent fitness costs of alarm signaling in aphids. Can J Zool 81:757–762
Montgomery ME, Nault LR (1977) Comparative response of aphids to the alarm pheromone, (E)-beta-farnesene. Entomol Exp Appl 22:236–242
Nakamuta K (1991) Aphid alarm pheromone component, (E)-beta-farnesene, and local search by a predatory lady beetle, *Coccinella septempunctata* bruckii Mulsant (Coleoptera: Coccinellidae). Appl Entomol Zool 26:1–7
Nault LR, Edwards LJ, Styer WE (1973) Aphid alarm pheromones: secretion and reception. Environ Entomol 2:101–105
Nault LR, Montgomery ME, Bowers WS (1976) Ant-aphid association – role of aphid alarm pheromone. Science 192:1349–1351
Pickett JA, Griffiths DC (1980) Composition of aphid alarm pheromones. J Chem Ecol 6:349–359
Pinto D, Blande J, Nykänen R, Dong W, Nerg A, Holopainen J (2007) Ozone degrades common herbivore-induced plant volatiles: does this affect herbivore prey location by predators and parasitoids? J Chem Ecol 33:683–694
Roitberg BD, Myers JH (1979) Behavioural and physiological adaptations of pea aphids (Homoptera: Aphididae) to high ground temperatures and predator disturbance. Can Entomol 111:515–519
Roscher C, Temperton VM, Scherer-Lorenzen M, Schmitz M, Schumacher J, Schmid B, Buchmann N, Weisser WW, Schulze ED (2005) Overyielding in experimental grassland communities – irrespective of species pool or spatial scale. Ecol Lett 8:576–577
Stadler B, Dixon AFG, Kindlmann P (2002) Relative fitness of aphids: the effect of plant quality and ants. Ecol Lett 5:216–222
Stadler B, Fiedler K, Kawecki T, Weisser WW (2001) Costs and benefits for phytophagous myrmecophiles: when ants are not always available. Oikos 92:467–478
Stadler B, Kindlmann P, Smilauer P, Fiedler K (2003) A comparative analysis of morphological and ecological characters of European aphids and lycaenids in relation to ant attendance. Oecologia 135:422–430

Stadler B, Weisser WW, Houston AI (1994) Defense reactions in aphids – the influence of state and future reproductive success. J Anim Ecol 63:419–430

Strauss SY, Rudgers JA, Lau JA, Irwin RE (2002) Direct and ecological costs of resistance to herbivory. Trends Ecol Evol 17:278–285

Tian D, Traw MB, Chen JQ, Kreitman M, Bergelson J (2003) Fitness costs of R-gene-mediated resistance in *Arabidopsis thaliana*. Nature 423:74–77

Tollrian R, Harvell CD (eds) (1998) The ecology and evolution of inducible defences. Princeton University Press, Princeton

Venables WN, Ripley BD (1997) Modern applied statistics with S-plus, 2nd edn. Springer, New York

Weisser WW (2000) A host-parasitoid metapopulation. Entomol Exp Appl 97:83–92

Wohlers P (1980) Die Fluchtaktion der Erbenlaus *Acyrthosiphon pisum* ausgelöst durch Alarmpheromon und zusätzliche Reize. Entomol Exp Appl 27:156–168

Yuan J, Himanen S, Holopainen J, Chen F, Stewart C (2009) Smelling global climate change: mitigation of function for plant volatile organic compounds. Trends Ecol Evol 24:323–331

Zhu JW, Cosse AA, Obrycki JJ, Boo KS, Baker TC (1999) Olfactory reactions of the twelve-spotted lady beetle, *Coleomegilla maculata* and the green lacewing, *Chrysoperla carnea* to semiochemicals released from their prey or host plant: electroantennogram and behavioural responses. J Chem Ecol 25:1163–1177

Species Index

P. Kindlmann et al. (eds.), *Aphid Biodiversity under Environmental Change*, DOI 10.1007/978-90-481-8601-3, © Springer Science+Business Media B.V. 2010

Subject Index

The manufacturer's authorised representative in the EU is Springer Nature Customer Service Centre GmbH, Europaplatz 3, 69115 Heidelberg, Germany. If you have any concerns regarding our products, please contact ProductSafety@springernature.com

Printed and bound by CPI Group (UK) Ltd, Croydon, CR0 4YY
25/09/2026
02222898-0004